THE ROLE OF THE
CHEMIST
IN
AUTOMOTIVE
DESIGN

THE ROLE OF THE
CHEMIST
IN
AUTOMOTIVE
DESIGN

H.K. PHLEGM

CRC Press
Taylor & Francis Group
Boca Raton London New York

CRC Press is an imprint of the
Taylor & Francis Group, an **informa** business

CRC Press
Taylor & Francis Group
6000 Broken Sound Parkway NW, Suite 300
Boca Raton, FL 33487-2742

First issued in paperback 2017

ISBN 13: 978-1-138-11426-5 (pbk)
ISBN 13: 978-1-4200-7188-7 (hbk)

Library of Congress Cataloging-in-Publication Data

Phlegm, H. K. (Herman K.)
      The role of the chemist in automotive design / author, H.K. Phlegm.
            p. cm.
      "A CRC title."
      Includes bibliographical references and index.
      ISBN 978-1-4200-7188-7 (hardcover : alk. paper)
      1. Automobiles--Materials. 2. Chemistry, Technical. 3. Automobiles--Design and
construction. I. Title.

TL240.P52 2009
629.2'32--dc22   2009013571

**Visit the Taylor & Francis Web site at
http://www.taylorandfrancis.com**

**and the CRC Press Web site at
http://www.crcpress.com**

To my daughters, Barbara and Lynn Phlegm;
my mother, Ollie Phlegm; and Tina Rae

Contents

Preface

The chemist's role in the automotive industry today is varied and a large number of chemists are being utilized outside a laboratory role. Many engineering applications require material knowledge above and beyond that of mechanical engineers. Chemists have been filling this role quite well as material engineers and product engineers. Polymers are increasingly used in today's vehicles in an effort to make them lighter, more cost effective, and more environmentally friendly. New technologies such as engineering polymers, polymer films in fuel cells, and CO_2 AC systems continue to drive this need.

I noticed the need for a book such as this through my guest lectures at the University of Detroit-Mercy and my experiences dealing with students and interns at General Motors. Their questions regarding what a chemist's role should be within the automobile industry prompted me to write this book. It is intended to give an overview of topics dealt with by today's automotive chemist; however, for more detail I recommend further reading.

Many circumstances involving the various duties of the chemist, mechanical engineer, and metallurgist have come to my attention over the last 20 years. Engineering polymers such as polyaryletherketones and polyimides provide better PV (pressure, velocity) characteristics, coefficient of friction (μ), and mass properties than metallic seals of the past. Proton-exchange membrane fuel cells use thin solid membranes in their construction. These membranes are typically a copolymer of tetrafluoroethylene and perfluorinated monomers. Microcellular foam technology involves an introduction of small amounts of supercritical nitrogen gas into molten polymer resin. This technology helps remove stress from parts as well as saves manufacturing cost. CO_2 air conditioning systems allow for the replacement of environmentally unfriendly fluorinated hydrocarbons. All of these technologies require the technical knowledge of chemists but also cross the lines of the various disciplines.

I would have liked to discuss many other topics, such as tire chemistry, the technology of adhesion, and various other issues important to the automobile industry. However, due to various limitations, I hope to address these topics at another time. I have tried to focus on what has been most relevant within my knowledge and could be helpful to the general layperson or student.

Herman K. Phlegm

The Author

Herman Phlegm was born in Pontiac, Michigan. He earned a master's degree in chemistry from the University of Detroit-Mercy in Detroit, Michigan. He attended undergraduate school at Texas Southern University and the University of Houston, Texas, where he studied chemistry and mathematics.

The author is currently employed at the General Motors Corporation. Over the last 23 years he has filled several chemistry-related positions at General Motors: laboratory chemist, chemical process engineer, seal design engineer, HVAC engineer, and alternate propulsion engineer. Prior to this, he was a water department chemist for the city of Detroit. He has several patents pending, as well as a defensive publication. Phlegm is the former chair of the Minority Affairs Committee for the Detroit branch of the American Chemical Society. He lives in Oak Park, Michigan with his two children.

1 Introduction to the Automobile Industry

1.1 INTRODUCTION

Each corporation in the automobile industry must face many challenges in order to remain competitive in today's environment. To be successful, an automobile manufacturer must react quickly to a changing market on a predictive basis. Paramount to this goal is the ability to utilize resources effectively. When we think of resources within a corporation, we generally think in terms of facilities, equipment, labor, capital, and human resources. In this text we want to concentrate specifically on human resources and, within that topic, the chemist's role within engineering resources. This role has been traditionally defined as a chemical analyst working within a major OEM (original equipment manufacturer) doing quality control or failure analysis. Much of the development work has been left to mechanical engineers as well as metallurgists. From the design and development of a vehicle to its manufacture and testing, it has become increasingly clear that chemical knowledge is required.

In this chapter I present some well known historic factors of past events and how they affect us today. In addition, I present a simple model of how the automobile industry develops and designs a product and the opportunities present for a chemist's involvement.

1.2 HISTORICAL FACTORS AFFECTING TODAY'S INDUSTRY

From a period of around the Yom Kippur War (1973) through the onset of the gasoline crisis it became apparent that things must change within the automotive industry. OPEC (Organization of Arab Petroleum Exporting Countries) supported the coalition of Arab states who were fighting against Israel in the Yom Kippur War. In preparation for the war, Saudi King Faisal and Egyptian President Anwar Sadat met in Riyadh and secretly negotiated an accord whereby the Arabs would use the "oil weapon" as part of the upcoming military conflict [1]. This meant that they would no longer ship petroleum to countries that supported Israel in the war. This resulted in an oil embargo on the United States. The embargo had an inflationary effect, with prices rising dramatically due to the decrease in the supply of oil. The price of oil quadrupled by 1974 to nearly $12 per barrel [2]. As a reference, the price of oil per barrel in 2008 hit a high of $145.95.

Up until that time, product portfolios within the big three automakers (Ford, Chrysler, and GM) had a typical averaged horsepower rating (for a 1972 mid-sized coupe) of 250–300 bhp with vehicle masses up to 3,900 lb, depending on the options.

Much of the vehicle mass was made up of metallic components that were later replaced by composites (e.g., intake manifolds, rocker covers, brackets). In addition, the efficiencies of these engines were not very good and needed to be improved.

Another factor facing U.S. automakers was foreign competition, specifically that from Japan. In 1964, Japanese four-wheeled motor vehicle exports were just over 100,000 [3]. However, according to the Japan Automobile Manufacturers Association (JAMA), in 1965 a 51.9% share of the export market in 1965 rose steadily to a 73% share in 1971. Passenger car exports in Japan's total automobile production output rose from 10.4% in 1965 to 40% in 1974; unit figures rose from 100,000 in 1965 to 1,827,000 in 1975 [4].

1.3 COMPETITIVE IMPERATIVES

These factors brought the collective leadership to recognize within the industry the need to remain competitive and to improve vehicles on a fuel economy basis. Changes to the cost structure while maintaining performance were needed as well. Eventually, this drive would manifest itself in the form of design imperatives of which the chemist would be crucial to development. From a management standpoint, these imperatives are

- performance;
- mass;
- cost; and
- environment.

As we will see in Chapter 4, each one of these imperatives is intermingled within the overall structure of the industry and can be considered to play a major role in competitive design. Because design plays such a major role in the overall cost structure of a vehicle (85%), obviously getting it right up front is crucial.

1.4 INDIFFERENCE MAPS AND CURVES

From a business standpoint, the marketing department will typically determine the direction in which it feels consumer trends are headed and tests several concepts in its attempts to do this. Economists must first make certain assumptions, such as transitivity, indifference, and utility. Marketing must decide whether a consumer who prefers a GM to a Ford, but a Ford over a Chrysler, would prefer a GM over a Chrysler. They must also determine if the customer is indifferent toward certain brands. If one is indifferent to the choice between Mercedes Benz and Subaru, for example, then one must also be indifferent to the choice between Volvo and Mercedes. The auto industry marketing group also assumes that a customer always prefers more over less. For example, if a customer prefers a vehicle package with 177 bhp, then he will choose a vehicle package with 260 bhp if given the opportunity. Obviously, we see where cost can come into play. In order to deal with this, we can group certain vehicle packages together. The indifference curve attempts to represent a consumer's taste in a manner that groups vehicle packages together to the point at which the consumer is indifferent.

TABLE 1.1
Vehicle Option Package and Attributes

Vehicle Option Package	Towing Capacity (In Pounds)	Feature Comfort Option Level
1	13,000	13
2	13,000	10
3	17,000	8
4	17,000	6
5	19,000	4
6	19,000	2
7	21,000	2
8	21,000	1

To take a fictional example, in Table 1.1 we examine eight option packages within the same platform (i.e., same frame and vehicle family) on a truck application. Obviously, many other factors would come into play, but for simplicity's sake we will examine only towing capacity and comfort package options. These comfort options include things in a package such as antilock braking, power seats, heated seats, automatic air conditioning, and luxury interior. Towing capacity was chosen due to its effect on the power train, which in turn affects fuel economy. The towing capacity will also affect the front end airflow, which in turn affects the styling of fascia design and coefficient of drag. Suppose now that we plot vehicle options as a function of towing capacity versus feature comfort level in Figure 1.1.

This curve is called an indifference curve. Higher levels of satisfaction can be realized by manipulating these curves by offering more of one feature or another and altering the tremendous number of variables in which people are interested (e.g.,

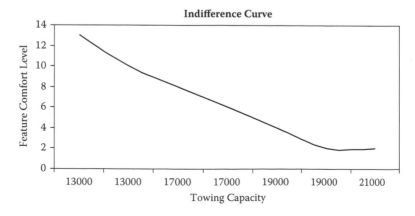

FIGURE 1.1 Indifference curve of comfort level versus towing capacity.

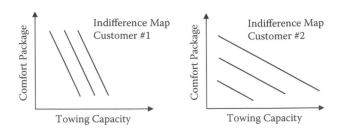

FIGURE 1.2   Indifference maps of two different customers.

tire pressure monitoring systems, chrome wheels, or leather seating). Now, if we imagine these curves in a series, we will have what is known as an indifference map. According to Mansfield, "A consumer's indifference map lies at the heart of the theory of consumer behavior, since such a map provides a representation of the consumer's taste" [5]. The shape of these indifference maps gives the marketing group an indication of the options and desires of the customer.

Figure 1.2 shows the indifference maps of two different customers. The slope of customer 1 indicates a greater level of satisfaction with towing capacity than that of customer 2. It is tools such as the indifference maps that allow the marketing groups to make recommendations to a program team. What is known as a "utility" is given to a comfort package or vehicle package. The utility is a number associated with a particular package. Mansfield indicates that the utility is the level of enjoyment or preference attached by this customer to a particular package [5]. Each curve, then, yields that same utility. Looking at different curves and packages will give an indication of what will be preferred by the customer.

1.5   MARKET DEMAND

While he was chairman of General Motors Corporation, John Smith was much more concerned with the quantity of cars that would be demanded by the entire national market than with the quantity of cars that an individual would purchase in the next year [5]. Part of the marketing task is to attempt to predict what this quantity demanded by the nation will be. The demand side is usually represented by a market demand schedule or table that shows the quantity of goods that will be purchased at a particular price. From the market demand table, a market demand curve is established. This is basically a price in some currency versus the quantity demanded in millions of units per year. (Again, Figure 1.3 is used for demonstration only.)

The downward slope of the curve indicates that demand increases as the price falls. However, this curve is shifted by certain factors such as taste of the customer. For instance, if the environmental imperative is legislated to the automobile manufacturer and the increasing desire to go green is present, then the demand will shift to the left. If the customer's awareness around global warming and environmental concerns manifests itself by a preference for a hybrid or electric vehicle over a 13-mpg, SUV, the market demand will shift as in Figure 1.4.

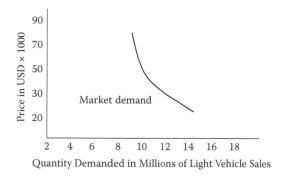

FIGURE 1.3   Market demand curve.

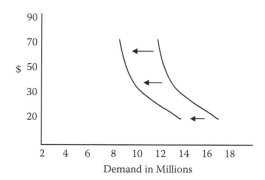

FIGURE 1.4   Market demand shift.

The demand curve is also affected by the level of a consumer's income. In the early 2000s, some industry analysts were predicting this demand shift would occur for the SUV market. However, sales continued to escalate, partly due to slight increases in the median annual household income. In 2006, the median value was $48,201, according to the U.S. census bureau [6]. By contrast, in 2000, the median value was $41,994. This may or may not have been a factor; however, the curve was shifted slightly to the right as demonstrated by increased vehicle sales during this period. Many things must be considered when such an analysis is made. Another factor that affects the demand curve is the level of other prices. If the price of natural gas and other necessary goods increases, then the demand for new vehicles will be curtailed.

Once the market demand has been established, a program team can set targets for a particular vehicle line. Depending on the company, targets can be established anywhere from 3 to 4 years. Typically, 3 to 3.5 years will allow a company adequate time to execute a new program. Architecture must be approved, designs must be detailed, engineering solutions applied, performance requirements set, etc. A great many tasks take place while a vehicle is being developed that are too detailed and numerous to discuss here. However, some of the associated tasks are listed in Table 1.2 to demonstrate where chemist involvement becomes crucial to the design

TABLE 1.2
Vehicle Development Tasks and Subtasks

Market determination
Vehicle performance targets
        Vehicle mass targets[a]
        Associated power train cooling requirements[a]
        Vehicle speed
        Towing requirements (if any)[a]
        Associated power train cooling requirements[a]
        Comfort requirements
        Air conditioning[a]
        Heating[a]
        Windshield clearance/defrost[a]
Environmental targets
    CO_2 emissions[a]
        MPG targets[a]
Architecture development
Concept development
Program framing
Long lead item development (sheet metal, HVAC, power train, frame, etc.)[a]
Electrical
Interior trim development[a]
Math data analysis
Vehicle builds
        Mules
        Integration vehicles
        Manufacturing builds
Supplier quality work
Start of production

[a] Task requires input from chemist.

and development of a vehicle. I will briefly describe the chemist's role in each of the vehicle development tasks listed and go into greater detail later in the book.

1.6 VEHICLE MASS TARGETS

Targets, including mass, are established early in a vehicle program. Typically, a vehicle mass group exists within an OEM; members of the group examine each aspect of the vehicle, system by system and part by part. They will specifically work with the engineer responsible for its design and, together with the engineer, will attempt mass reduction. This can take many forms. For instance, it can be in design by adding a stiffening rib while removing more mass in other areas. Within the past 15 years, finite element analysis has aided in this process a great deal and is integral in the development of a part. Another technique is simply material substitution—that is, substituting a lower density approved material for what is currently used. In either case, the process involves looking at the material strength required, tensile as well as compressive.

The material density must be verified and, depending on the application, chemical resistance and thermal properties examined. In general, for each part within a sub-group of a vehicle system, the material engineers have a designated list of approved materials from which the design engineer can choose. This is not to say that a new material cannot be introduced for a new program application; rather, this list provides a basis from which the design engineer can start.

1.7   POWER TRAIN COOLING REQUIREMENTS

The following are the major factors that determine the power train cooling requirements for a vehicle:

- vehicle mass;
- type of engine;
- engine heat rejection ability;
- transmission;
- transmission gear ratio;
- engine and transmission calibrations;
- vehicle axle diameter;
- AC system used;
- towing or load-bearing requirements; and
- coefficient of drag for the vehicle.

Chemist input here is through the HVAC (heating, ventilation, and cooling group). Input is typically the material selection and processing within a heat exchanger that will give the best heat transfer characteristics (i.e., radiator, heater core, evaporator core, condenser, alternate coolers). A chemist working as material engineer has input to the transmission oil and its capacity for cooling: Is the transmission oil optimized for the best performance? What are its wear characteristics? What is the anticorrosion package like and is it compatible with the transmission parts? Likewise, with the cooling requirements for the engine, the chemist input involves the type of oil chosen (synthetic or mineral), the associated additive packages, and the wear characteristics, as well as such things as oxidation resistance, and seal compatibility.

The size of a cooling system finally chosen may add stress to the system if it just barely meets requirements. New organic acid additive packages in engine coolant developed by chemists must maintain a minimum concentration in order to function properly. Their mechanism works by combining with the metal constituents of a vehicle versus simply coating the parts. These new additives are typically sebacic acid, 2-ethylhexanoic acid, or tolytriazole. All subsequent work with seal designs must be chemically compatible with the new types of additive packages.

1.8 HVAC

When a power-train cooling system or an AC system is designed, consideration must also be given to the refrigerant used by the system. A CO_2 system versus a 134a or 152a system will have a great effect on optimization of a design. CO_2 systems require extremely high operating pressures because CO_2 does not condense in the refrigeration circuit. We will discuss this in detail in Chapter 9. As a consequence, a cooler is used rather than a condenser. The reason for the use of CO_2 is that it does not impact global warming or ozone depletion. Different refrigerants have different heat transfer characteristics, subcooling and lubrication requirements, etc. A CO_2 system requires a secondary loop outside the vehicle compartment due to the danger of leakage within the cabin.

1.9 EMISSIONS

CO_2 comes into play not only from the refrigerant but also within the vehicle emissions. Obviously, we want to minimize what the engine is putting out in terms of emissions. In addition to CO_2, nitrogen gas (N_2), nitrogen oxide (NO_x), carbon monoxide, water vapor, and volatile organic compounds (VOCs) are produced. We attempt to get as near as possible to the stoichiometric point of gasoline to air of 14.7:1. This is done through mechanical means with engine electronics, exhaust gas recirculation (EGR) valves, electrical control, and, of course, catalytic converters.

Chemists are constantly working on ways to improve combustion and reduce the production of VOCs, not only for improving efficiency but also for reduction of loss. This could be as simple as replacing a steel fuel line and connection with that of a flexible nylon line that is less susceptible to leaks after consultation early in a program. Work is currently being done with three-stage converter systems by chemists working with fuel systems on catalytic material substitution. Typically, with a three-way converter utilizing a reduction and oxidation catalyst, we have a ceramic structure coated with platinum, palladium, or rhodium catalyst. The trick here is to have enough surface area to expose to the exhaust without increasing the amount of the extremely expensive catalyst used. The reduction catalyst will reduce the NO_x emissions by removing the nitrogen atom from the molecule and combining it with other nitrogen atoms, thus leaving a molecule of nitrogen and a molecule of oxygen. Equation 1.1 represents this process.

Reduction reaction:

$$2NO \rightarrow N_2 + O_2$$

$$2NO_2 \rightarrow N_2 + O_2 \qquad (1.1)$$

The oxidation catalyst reduces the amount of unburned VOCs and CO by oxidation at the Pt and Pd catalysts. Equation 1.2 represents this process.

Oxidation reaction

$$2CO + O_2 \rightarrow 2CO_2 \tag{1.2}$$

Current research centers on utilizing gold as a catalyst due to the extremely high cost of the current Pt and Pd catalyst systems. Chemists have recently developed a carbamide (urea) injection system with diesel engines. Diesels do not reduce NO_x emissions as effectively as gasoline engines due to their lower operating temperature. With these systems, a carbamide solution is injected into the exhaust pipe before being injected into the catalytic converter. The carbamide injection reacts with NO_x to produce nitrogen and water vapor. Design is where the most vehicle cost impact can be made, and taking advantage of the best system and material choices in the design phases will generate material savings by up to 85%.

1.10   GREEN ALTERNATIVES

In today's market, vehicle teams attempt to offer a green alternative in their vehicle portfolio. This will typically be a hybrid vehicle or electric vehicle, with fuel cell options soon to come. If a hybrid option is chosen, development work has been greatly influenced by the corporation's chemist. Hybrids have a rechargeable energy storage system (RESS) as well as a traditional engine. With an RESS, the electric battery is enhanced by an electrochemical double layer capacitor (EDLC). These capacitors are nearly 1,000 times greater in capacitance than a typical capacitor. These types of capacitors were derived from Helmholtz's model of the electrical double layer. The benefits to the environmental imperative are obvious. The power train's internal combustion engine size can be reduced. Energy can be recaptured with things such as regenerative breaking, and energy can be reduced by not utilizing the engine at idle.

Just as important to these systems is the future use of fuel cells and fuel cell vehicles. Toyota and General Motors are currently utilizing chemists, chemical engineers, and mechanical engineers in the development of viable fuel cell vehicles. Hybrid electric vehicles and electric vehicles certainly help with the environmental imperative, although there are some problems. Electric vehicles have a range generally limited to around 100 miles. Batteries are large and require compartment space better utilized by other systems. They also hurt the mass imperative due to the greater mass requirements over a typical persistent bioaccumulative and toxic gasoline tank. Fuel cell vehicles have no moving parts and are not dependent on a nonrenewable fuel source. Hydrogen is a fuel source that is fully renewable. With these systems, a flat cell is separated by a membrane with a cathode and anode that form the cell. There is typically a platinum catalyst and an external circuit.

One of the major issues with fuel cells is the problem of hydrogen storage. A very large tank would be needed even with a compressed hydrogen tank. Other issues include carbon monoxide intolerance, overpotential at the anode and cathode, and internal resistance. We will be discussing these problems in greater detail in Chapter 12.

REFERENCES

1. Yergin, D. H. 1991. *The prize: The epic quest for oil, money, and power,* 597. New York: Simon and Schuster.
2. National Academy of Sciences. 1982. The competitive status of the U.S. auto industry in crisis; a study of the influences of technology in determining international industrial competitive advantage, 10 (http://books.nap.edu/openbook.php?record_id=291&page=10).
3. The price of oil—In context. Retrieved on May 29, 2007. http://www.oil-price.net/
4. Japan Automobile Manufactures Association. Japan's auto industry (http://www.jama.org/about/industry10htm)
5. Mansfield, E. 1997. Microeconomics, 51, 54, 110. New York: W. W. Norton & Company.
6. U.S. Census Bureau news release regarding median income (http://pubdb3.census.gov/macro/032007/hhinc/new04_001htm)

2 Traditional Role of the Chemist in the Automobile Plant Environment

2.1 INTRODUCTION

Within automobile management in a plant environment, traditional thinking led to the belief that a chemist's role was strictly within the laboratory in an analytical capacity or as support for some chemical system such as coolant (soluble oil), paint, or lubrication systems. In addition to this were the roles as analyst for a failed component and inspector in quality control for incoming raw materials. In this chapter, we will discuss these roles and briefly focus on some of the necessary theory and technology in fulfilling them. We will again see how the imperatives mentioned in Chapter 1 come into play within the auto industry and in the chemist's role, specifically.

During the period from roughly the 1930s to the late 1990s, auto companies depended on chemists to inspect incoming raw materials, maintain chemical processes, and perform failure analysis. These activities were generally done in the metallurgical (met) lab and ranged from wet methods to advanced instrumental techniques. It was not until later that the "just in time" inventory control system and reliance on supplier certification limited incoming chemical inspections. In this chapter, we will describe the chemist's task in analyzing the various materials and, in Chapter 3, will enter into more detail about the materials used.

2.2 INCOMING INSPECTION

For a typical production run, the materials inspection group would obtain a sample of five parts per lot. The entire lot of parts would then be quarantined and inventoried through the use of an automated system such as a FATA system. The samples from each lot would then be sent to the material lab for analysis. Depending on the nature of the material, the chemist would determine the type of testing required, complete the test, and compare it to the material specification. Table 2.1 shows typical equipment found in the automotive chemist's lab in the plant environment. This list is meant to be a general guide to typical equipment and is not meant to be inclusive.

Table 2.2 presents analytical signals. According to Skoog and West, these signals are common signals useful for analytical purposes [1]. Up until around 1920,

TABLE 2.1
Analytical Instrument Usage

Equipment	Primary Usage
Optical emissions spectrophotometer	Analysis of steel, iron, and aluminum components
HPLC (high-performance liquid chromatography)	Analysis of additives in coolants, oils, hydraulics
DSC-TGA (differential scanning calorimetry-thermal gravimetric analysis)	Analysis or polymeric components
AA (atomic absorption)	Analysis of steel, iron, and aluminum components
Carbon sulfur analyzer	Analysis of iron and steel components
IR and FTIR (Fourier transform infrared)	Analysis of manufacturing system chemicals as well as surface analysis
GC/MS (gas chromatography/mass spectrometry)	Material identification
Classical methods	General system analysis, paints, coolants, hydraulics, etc.

TABLE 2.2
Signal versus Analytical Method

Signal	Analytical Methods Based on Measurement of Signal
	Emissions spectroscopy (x-Ray, UV, visible, electron, auger)
Emission of radiation	Flame photometry; fluorescence (x-ray, UV, visible)
	Radiochemical methods
	Spectrophotometry (x-ray, UV, visible, IR); colorimetric; atomic
Adsorption of radiation	Absorption, nuclear magnetic resonance, and electron spin resonance
Scattering of radiation	Turbidimetry; nephelometry; Raman spectroscopy
Refraction of radiation	Refractometry; interferometry
Diffraction of radiation	X-ray and electron diffraction methods
Rotation of radiation	Polarimetry; optical rotatory dispersion; circular dichroism
Electrical potential	Potentiometer; chronopotentiometry
Electrical current	Polarography; amperometry; coulometer
Electrical resistance	Conductimetry
Mass to charge ratio	Mass spectrometry
Rate of reaction	Kinetic methods
Thermal properties	Thermal conductivity and enthalpy methods
Mass	Gravimetric analysis
Volume	Volumetric analysis

most analysis performed in the lab by automotive chemists was classical. Around this time, with the advent of electronics, equipment such as spectrophotometers began to make its way into the automotive lab. I will later talk more about chromatography because I believe it is an underutilized method in today's automobile lab. Chromatography is an instrumental analytical method used for separation and resolution of materials such as coolants and additive packages in hydraulic fluids that have closely related molecular structures. Distillation and extraction are classical methods used in the laboratory as separation techniques. These methods, however, are rarely used in today's labs; they are mostly preparation techniques for some other analytical method.

Depending on the sample and what type of analysis the chemist wants (qualitative or quantitative), the chemist will then conduct the analysis and report the results.

2.3 METHODS AROUND METALS

The primary components of automobiles are steel or aluminum, so one of the fastest methods for analysis with the least amount of preparation of the sample is the emissions spectrometer. From Table 2.1, we can see that a carbon sulfur analyzer, such as a Leco, or atomic absorption spectrophotometer; scanning electron microscopy (SEM); x-ray; and GC-MS are also used for this type of analysis. However, an emissions spectrophotometer is most often used because of its lack of sample preparation. Again, it is not our attempt here to go into great detail on each method. Within an automotive analytical laboratory, however, speed is a priority so that a material is identified and classified rapidly. An emissions spectrophotometer is such an instrument.

2.3.1 Emission Spectrophotometers

Generally, argon plasma sources have been used in this type of instrument since the late 1970s. The argon plasma source has greater energetic excitation than the atomic absorption discussed later. Electric arc and spark systems have been used in labs since the 1930s. Their popularity within the auto industry was partially because samples did not need to be prepared. Typical samples are prepared by simply grinding away the surface finish. Care must be taken, however, to make sure there is a flat area to rest on the sample stand of the spectrophotometer. Prior to analysis, the electrode (graphite for aluminum samples and steel for steel samples) must be gapped to the proper distance for analysis. Because of the higher energy involved, spectra for dozens of elements can be recorded simultaneously [1]. The excitation source must supply sufficient energy to vaporize the sample that is being analyzed. It must also be able to cause electronic excitation in the sample.

We are attempting to isolate line spectra from excited atoms and measure the intensity from these lines. To do this, we excite an element from the ground state to an excited state. For instance, a single outer electron from metallic sodium occupies the 3 s orbital at ground state. An electric spark or arc is used to excite the electron into a higher energy state. Upon return to the ground state, radiation is emitted at various wavelengths. The sodium energy level diagram is listed in Figure 2.1 as an example.

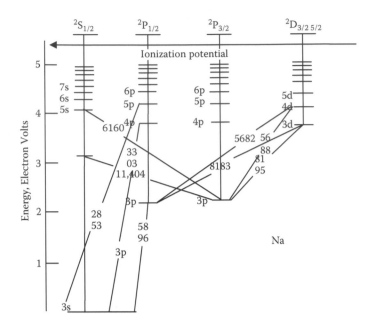

FIGURE 2.1 Sodium energy level diagram.

In addition to the line spectra produced by electric arc or spark, band spectra are produced as well as continuous background radiation. Band spectra come from the molecular species in the vapor state that is produced from the spark. The molecular energy levels are superimposed over the electronic energy levels. Typically, cyanogens (CN), siloxanes (SiO), and hydroxyl radicals are sources of band spectra. Continuous background spectra are produced from the arc and spark source itself. The heating of the electrodes will produce radiation that is temperature dependent and approximates that of a black body [1].

The most common type of emissions spectrometer used today is of the multielement type. This type of instrument has the speed advantage that the plant chemist wants as opposed to a sequential type of spectrophotometer. A multielement type can measure up to 60 elements at a time. When an arc or spark type of instrument is used, it is necessary to integrate and average the signal produced to obtain reproducible line intensities.

These types of instruments use photomultiplier tubes for each element. A schematic of this type of spectrometer is shown in Figure 2.2. Photomultiplier tubes are located behind exit slits along the focal curve of the instruments. These exit slits are placed at the corresponding position of the line being analyzed. After the sample is sparked, the deflection grating separates the emitted light, which then uses a Hartman diaphragm to mask a portion of this light. Separation of the emitted light generally falls into the spectral region between 2,500 and 4,000 Å. The photomultiplier tubes feed the current to a circuit called a capacitor–resistor circuit; this integrates the output of the tube. At the end of the cycle, the capacitor's voltage is a function of the total charge, which is related to the intensity of the line of the element being analyzed and thus its concentration.

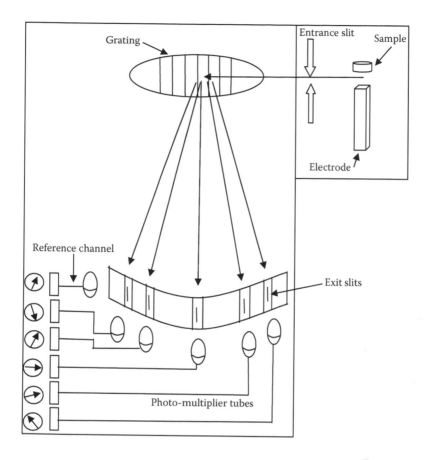

FIGURE 2.2 Diagram of emissions spectrophotometer.

These instruments are ideal for usage within the automobile chemist's laboratory due to their speed for routine examinations and their ability to analyze a multitude of elements with relatively high precision.

2.3.1.1 Detection Limits

Data from Spectrochim Acta show the wavelengths and lower detection limits for some elements usually found in metallic components used in automobiles (Table 2.3).

2.4 ATOMIC ABSORPTION FOR METAL ANALYSIS

Another useful and relatively rapid method used by chemists in the automobile chemistry lab is atomic absorption (AA). This technique has technically been around for quite some time; however, work done at the Commonwealth Science and Industry Research Organization led to the modern form of atomic absorption. Modern instruments can analyze as many as 60 elements. In AA, a flame or graphite furnace is used to atomize the sample. The sample is first placed in an aqueous solution and

TABLE 2.3

Detections Limits for DC ARC

| Element | Wavelength (Å) | Lower Limit of Detection | |
		Percent	Micrograms
Ag	3280.68	0.0001	0.01
As	3288.12	0.002	0.2
B	2497.73	0.0004	0.04
Ca	3933.66	0.0001	0.01
Cd	2288.01	0.001	0.1
Cu	3247.54	0.00008	0.008
K	3446.72	0.3	30
Mg	2852.12	0.00004	0.004
Na	5895.92	0.0001	0.01
P	2535.65	0.002	0.2
Pb	4057.82	0.0003	0.03
Si	2516.12	0.0002	0.02
Sr	3464.45	0.02	2
Ti	3372.8	0.001	0.1
Zn	3345.02	0.003	0.3

Source: Skoog, D. *Principles of Instrumental Analysis*, Saunders College, 1980. With permission.

aspirated into a hot flame or burner, where a substantial amount of the metal in solution is reduced to an elemental state. Monatomic ions, which were formed during the atomization [1], are also present. A schematic is presented in Figure 2.3.

The radiation source in AA is produced by a hollow cathode tube. A cathode tube must be utilized specifically to the metal being analyzed. The tube contains a cathode with the metal that is being analyzed and an anode across which a potential is applied thereby exciting electrons for that particular atom from the ground state into a higher energy state. The difference between the energy observed at the detector versus what was input can be calibrated and quantified, thereby giving the concentration of the element in question.

As in atomic emissions, some interferences from molecular species occur and must be corrected for. Molecular emissions contain much broader bandwidths and

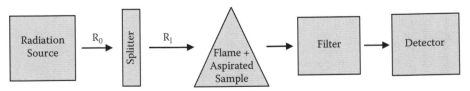

Typical Flame AA diagram. R_0 is from radiation source, R_I is radiation at λ of interest

FIGURE 2.3   Diagram of flame AA.

can overlap the narrow lines from a hollow cathode tube. Corrections such as Smith–Hieftje have been employed [2].

2.5   SEPARATION AND CHROMATOGRAPHY OF ORGANICS

Separation techniques can be very useful when dealing with plant fluid systems such as coolant cutting fluids, paints, and hydraulic fluids including systems such as transmission fluids. One of the most useful techniques in my experience in the automobile laboratory was chromatography. Traditionally, there are many different types of chromatography, all of which employ a stationary phase and a mobile phase. The compounds to be analyzed (analytes) will be carried by the mobile phase and pass through a stationary phase. Separation is achieved by the different rates at which the sample goes through the stationary phase. I will not go into a detailed discussion of the various types of chromatography here, but rather try to focus on what is most useful for today's automobile industry chemist.

2.6   LIQUID–SOLID ADSORPTION IN HPLC

Liquid–solid or adsorption chromatography, used in high-performance liquid chromatography (HPLC), is a very useful laboratory tool in analyzing many systems in the automobile plant. For instance, in a plant environment, various machining operations take place, such as machining of aluminum, steel, or some other type of alloy. During these processes, vast amounts of heat are generated and require cooling from a cutting fluid. The machining operation could be from cutting, grinding, milling, turning, etc. Most automobile manufacturers deal with water-soluble cutting fluids, which contain a base oil and emulsifying agent, various biocides, antiwear compounds, lubricity compounds, and surfactants. Some of these compounds lend themselves to analysis by HPLC.

2.7   SOLUBLE OILS

In addition to water-soluble fluids, there are synthetic and semisynthetic fluids. According to Watanabe [3], the JIS (Japanese Standards Association) has three classes of water-soluble fluids. Type A1 (emulsion type) contains a base oil and the emulsifying agent and clouds when mixed with water. Type A2 (soluble type) contains surface active agents and is translucent when mixed. Type A3 is a chemical solution type; it contains organic and inorganic carboxylic acids and is also translucent. Type A2 is most often used in automobile manufacturing plants.

When these types of cutting fluids are manufactured, they typically will contain an ethanolamine salt or some form of emulsifying agent that is used to get the oil in solution. Many of the best formulations will contain triethanolamine salts because of their excellent ability to address what the chemist is looking for—typically, lubricity and anticorrosion protection. Typical ethanolamine salts of *t*-butylbenzoic acid, pentylbenzoic acid, hexylbenzoic acid, and *p*-butoxybenzoic acid have been chosen due to their favorable characteristics from the standpoint of cost, solubility, corrosion resistance, and load ability [3].

Some of these carboxylic acid salts, such as 3-iodo-4-methosxybenzoic acid, have been reported to have load-bearing capacities in an aqueous solution of 0.98 (megapascal) MPa at 200 rpm with a four-ball type lubrication oil testing machine [4]. A four-ball wear test (ASTM D-4172) is a test used in the petroleum industry that determines the wear characteristics of a lubricant. In this test, three balls are clamped together, covered with the lubricant being analyzed, and rotated with a fourth ball under a load. Inputs can be varied by temperature, pressure, revolutions per minute, and duration.

2.8   LUBRICITY ADDITIVES

In addition to carboxylic acid salts, other effective lubricity additives that the chemist will work with are esters of hydroxyfatty acids, non-ionic surface agents, dibasic acids, and additives from substituted fatty acids. Although this is not a complete list, its contents are the most effective from a cost perspective and are typically used in the industry. The oleate of ricinoleic acid dimer is an example of a hydroxyfatty acid prepared from ricinoleic acid oligomer with acid chloride in the presence of pyridine, as shown in Figure 2.4.

These compounds make excellent lubricity additives. It is possible that rotation about the oxygen atom in the ether linkage allows for the molecule to bend back upon itself, thus providing greater flexibility within the chain. Other preparations have been reported by Watanabe in the trimer, tetramer, and hexamer variety [4]. Watanabe and colleagues have also shown that esterifying the hydroxyl group of block copolymers of non-ionic surfactants (Figure 2.5) gives good antiwear properties of the species synthesized.

These esters were prepared from aqueous solutions of tri-ethanolamine and PE-61.PE-62 and PE-64 block copolymers. When higher monoesters of the surface active agents were utilized, excellent antiwear properties were observed. Again, the configuration of the ether linkage in relationship to the rest of the molecule plays a role in this compound's effectiveness. Table 2.4 shows properties of cutting fluids from ricinoleic acid oligomer esters with triethanolamine. The same values from fluids made from polymeric non-ionic surfactants are shown in Table 2.5.

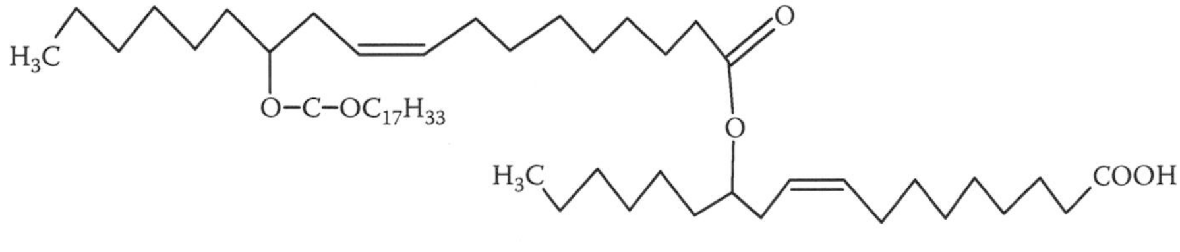

FIGURE 2.4   Structure of oleate of ricinoleic acid.

$$HO-(C_2H_4O)_n-(C_3H_6O)_m-(C_2H_4O)_n \xrightarrow[\text{2. RCOCl}]{\text{1. Pyridine}} R-\underset{O}{\overset{\parallel}{C}}-O-(C_2H_4O)_n-(C_3H_6O)_m-(C_2H_4O)_n$$

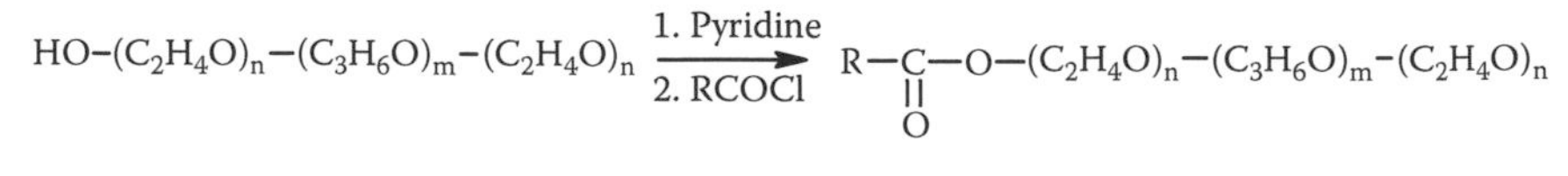

FIGURE 2.5   Non-ionic surfactant.

TABLE 2.4

Property Table of Salts of Ricinoleic Acid Oligomer Cutting Fluids

R or Esters of Polymer (X) (X-OCOR)	pH	Rust Inhibition Test Duration (h)[a]			Friction Coefficient	Surface Tension 10^{-3} (N/m)	Initial Seizure Load (MPa)
		24	48	72			
Dimer-O-COR							
C_9H_{19}	9.0	10	10	10	0.09	41	1.91
$C_{11}H_{23}$	9.1	10	10	10	0.08	41	1.76
$C_{17}H_{33}$	9.4	10	10	10	0.08	41	1.96
Trimer-O-COR							
C_9H_{19}	9.0	10	10	10	0.09	41	1.96
$C_{11}H_{23}$	8.9	10	10	10	0.08	42	1.96
$C_{17}H_{33}$	9.3	10	10	10	0.08	43	1.76
Tetramer-O-COR							
C_9H_{19}	9.0	10	10	10	0.09	43	1.96
$C_{11}H_{23}$	9.1	10	10	10	0.08	44	1.96
$C_{17}H_{33}$	9.6	10	10	10	0.08	41	1.96
Hexamer-O-COR							
C_9H_{19}	9.4	10	10	9	0.09	44	1.96
$C_{11}H_{23}$	9.3	10	10	10	0.08	45	1.96
$C_{17}H_{33}$	9.3	10	10	10	0.08	37	1.96
Ricinoleic acid	8.8	10	9	8	0.12	43	1.81
Ricinoleic acid dimer	9.1	10	9	8	0.1	41	1.86
Ricinoleic acid trimer	9.7	9	9	7	0.09	37	1.67
Ricinoleic acid tetramer	9.8	9	9	7	0.09	36	1.81
Ricinoleic acid hexamer	10.0	9	9	7	0.09	43	1.91
Triethanolamine (2% aqueous solution)	10.1	7	6	5	0.21	70	0.83

[a] Evaluation of rust inhibition test:

Amount of Rust	Valuation Point
No appearance of rust	10
One point of rust	9
Two points of rust	8
Some points of rust	7
Many points of rust	6
Many points of rust and stains	5

Source: Watanabe, S. *J. Oleo Sci.*, 56, 385, 2007. With permission.

TABLE 2.5
Property Table of Esters of Non-ionic Surface Active Agents

Esters of Polymers	pH	Rust Inhibition Test Duration (h)[a]			Friction Coefficient	Surface Tension 10⁻³ (N/m)	Initial Seizure Load (MPa)
		24	48	72			
PE-61 Monoester							
PE-61	9.5	8	7	6	0.17	38	0.54
Acetate	9.1	8	8	7	0.14	39	1.81
Butyrate	9.3	8	8	7	0.19	39	1.22
Hexanoate	9.4	10	10	9	0.14	39	1.37
Octanoate	9.5	10	10	10	0.18	38	0.74
Decanoate	9.8	10	10	9	0.16	37	1.47
Oleate	9.6	10	10	10	0.15	38	0.83
p-t-Bentylbenzoate	9.7	10	10	10	0.13	38	0.83
PE-61 Diester							
Diacetate	9.6	8	7	7	0.1	38	0.78
Dioctanoate	9.6	10	10	10	0.18	39	1.08
Dioleate	9.6	10	10	10	0.13	38	1.03
PE-62 Monoester							
PE-62	9.9	8	7	6	0.2	38	0.88
Acetate	9.2	8	6	6	0.22	38	0.93
Oleate	9.6	10	10	10	0.2	37	0.98
p-t-Bentylbenzoate	9.3	10	10	10	0.23	37	0.78
PE-62 Diester							
Diacetate	8.8	8	7	7	0.18	37	0.88
Dioleate	9.6	10	10	10	0.13	38	1.96
p-t-Bentylbenzoate	9.4	10	10	10	0.22	37	0.88
triethanolamine (2% aqueous solution)	10.1	7	6	5	0.26	70	0.54

[a] Evaluation of rust inhibition test:

Amount of Rust	Valuation Point
No appearance of rust	10
One point of rust	9
Two points of rust	8
Some points of rust	7
Many points of rust	6
Many points of rust and stains	5

Source: Watanabe, S. *J. Oleo Sci.*, 56, 385, 2007. With permission.

The chemist in the automobile industry may be called upon not only for the manufacture of these compounds but also for the analysis in order to maintain the proper concentration of the compound. The detection in an HPLC system is most widely accomplished by UV-visible detectors. Most compounds used in production chemicals have chromophores that are detectable in the UV region. Table 2.6 shows molar absorption values for common functional groups [5]. Care must be taken that the solvent used in the analysis is invisible to the analysis. As can be seen from Table 2.6, typical detection range is 180–210 nm. A chromophore is the region between two different molecular orbitals and will absorb UV light due to the conjugated system of electrons or to a metal complex bound to a transition metal to a ligand.

2.9 SOME PROBLEMS WITH HPLC AS A LAB TOOL

In HPLC, an analyte is forced through a column of a stationary phase utilizing high pressure. A sample of the chemical being analyzed is then introduced into the mobile phase. Within the stationary phase, the sample will interact with the column material and be separated within the column by chemical and physical interactions. The variables of the sample itself, the column, and the mobile phase all contribute to the retention time on the column. As the separation of the sample occurs within the column, the pressure of the HPLC will increase the speed and reduce the diffusion seen in a normal chromatogram. Typically, methanol and acetonitrile are used as the solvent in the mobile phase for separations dealing with cutting fluids and hydraulic additives. Most often a gradient will be applied to vary the concentration of methanol/acetonitrile to enhance the separation.

Reverse phase HPLC is also used due to the nature of the polar constituents that are examined. In reverse phase, the stationary phase is nonpolar and the mobile phase will be a mixture of polar solvents and aqueous media.

Silica columns are generally used with long chain alkyl groups connected to $R(CH_3)_2SiCl$. The polar fats, amines, and other polar functional groups in the additive packages of concern will then come through the column more readily than the longer chain nonpolar components in the base oils. From the discussion on lubricity additives, one can see that reverse phase chromatography is a good choice for separation of these compounds.

2.10 PLATE THEORY AND RATE THEORY

Martin and Synge first described plate theory in chromatography in the early 1940s [6]. Their thought was that a column can be considered as a long continuum of narrow horizontal layers called theoretical plates. These theoretical plates are discrete but thought of as being contiguous. Between the plates, equilibrium with the solute and the two phases takes place. Thinking of a separation this way envisions a stepwise process between each theoretical plate. A useful term, "height equivalent to theoretical plate, H," is related to the efficiency of a column by the equation:

TABLE 2.6

Molar Absorbtivity of Some Functional Groups

Compound Type	Chromophore	Wavelength (nm)	Molar Absorptivity
Acetylide	$-C{\equiv}C-$	175–180	6,000
Aldehyde	$-CHO$	210	1,500
		280–300	11–18
Amine	NH_2	195	2,800
Azido	$C{=}N$	190	5,000
Azo	$-N{=}N$	285–400	3–25
Bromide	$-Br$	208	300
Carboxyl	$-COOH$	200–210	50–70
Disulfide	$-S{-}S$	194	5,500
		255	400
Ester	$-COOR$	205	50
Ether	$-O-$	185	1,000
Iodide	$-I-$	260	400
Ketone	$C{=}O$	195	1,000
		270–285	15–30
Nitrate	$-ONO_2$	270	12
Nitrile	$-C{\equiv}N$	160	—
Nitrite	$-ONO$	220–230	1,000–2,000
		300–400	10
Nitro	$-NO_2$	210	Strong
Nitroso	$-N{=}O$	302	100
Oxime	$-NOH$	190	5,000
Sulfone	SO_2	180	—
Sulfoxide	$\overset{H}{\underset{H}{\diagdown}}S{=}O$	210	1,500
Thioether	$-S{-}O$	194	4,600
		215	1,600
Thioketone	$C{=}S$	205	Strong
Thiol	$-SH$	195	1,400
Unsaturation, conjugated	$-(C{=}C)_3-$	260	35,000
	$-(C{=}C)_4-$	300	52,000
	$-(C{=}C)_5-$	330	118,000
Unsaturation, aliphatic	$-C{=}C-$	190	8,000
	$-(C{=}C)_2-$	210–230	21,000
Unsaturation, alicyclic	$-(C{=}C)_2-$	230–260	3000–8000
	$C{=}C{-}C{\equiv}C$	291	6,500
	$C{=}C{-}C{\equiv}N$	220	23,000
Miscellaneous compounds	$C{=}C{-}C{\equiv}O$	210–250	10,000–20,000
		300–350	Weak
	$C{=}C{-}NO_2$	229	9,500
Benzene		184	46,700
	C_6H_5	202	6,900
		255	170

Source: Poole, C. *Contemporary Practice of Chromatography,* Elsevier, 1984. With permission.

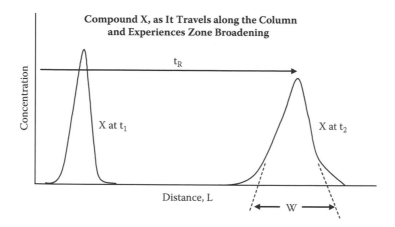

FIGURE 2.6 Zone broadening in HPLC.

$$N = \frac{L}{H} \tag{2.1}$$

Here, N is the number of theoretical plates and L is the length of the column. Plates do not actually exist, but rather are used merely as a consideration in column efficiency.

In a typical chromatogram, a fair amount of zone broadening takes place between solutes. Because of this, plate theory has been replaced by the theory of kinetics, or rate theory.

When zone broadening such as that shown in Figure 2.6 is present, plate theory is not adequate to describe it. In zone broadening, the plate height can be expressed in centimeters by taking the square of the variance per unit length of the column so that

$$H = \frac{\sigma^2}{L} \tag{2.2}$$

Alternatively, Equation 2.2 can be expressed as

$$N = \frac{L^2}{\sigma^2} \tag{2.3}$$

with σ having the units of centimeters. The time that the solute spends on the column is the retention time t_R. A value t_M (not shown here) is the time for the mobile phase to pass through the column.

The following equations are used by the plant chemist to calculate column efficiencies:

$$H = \frac{LW^2}{16t_R^2} \tag{2.4}$$

$$N = 16\left(\frac{t_R}{W}\right)^2 \qquad\qquad (2.5)$$

The calculation of N is derived from two measurements, t_R and W.

Skoog and West describe three causes of zone broadening: eddy diffusion, longitudinal diffusion, and nonequilibrium mass transfer [1]. The Van Deemter equation was developed to relate the flow rate and plate height:

$$H = A + \frac{B}{u} + Cu \qquad\qquad (2.6)$$

Eddy diffusion is represented by A and is simply the various paths that a molecule travels along the column before arriving at the detector. B is the quantity represented by longitudinal diffusion, which is the Gaussian type of migration from the center of a column toward the ends of a column. This factor is inversely proportional to the flow rate (u). Finally, nonequilibrium mass transfer has to do with the flow rate not being high enough to allow an equilibrium between the mobile and stationary phases. Nonequilibrium mass transfer becomes smaller as the flow rate is decreased [1].

There are so many important systems for the plant chemist to examine that HPLC must be considered as one of the primary methods for doing so. A good method is to examine the analyzable species in the compound and then determine the method of analysis and the other variables in the separation.

2.11   ELASTOMER CHARACTERIZATION

The use of rubber components in vehicles is extensive. Applications include sealing, vibration dampening, and trim. Most rubbers in power train applications are for seals and gaskets (Chapter 8). Chassis applications most often center around vibration dampeners (isolators) for heat exchangers, engine and transmission mounts, and exhaust line isolators. Body and interior applications include interior trim and body sealers. A more thorough list of applications and types of rubbers is presented in Chapter 3.

Sealing in an automobile is most vital and a chemist's skills are quite useful. Just as vital however, is the isolation and vibration dampening application of rubbers; thus, some discussion time should be devoted to this subject. Most mechanical engineers and computer analysts do not have a working knowledge of rubbers and therefore cannot make the material choices or computer predictions that the chemist can.

I should provide a brief explanation of some key terms when dealing with elastomers. Stress is the average amount of force exerted on an elastomer per unit area and can be represented by

$$\sigma = \frac{F}{A} \qquad\qquad (2.7)$$

Here, force is the quantity, F and A represents the unit area. The unit the plant chemist will usually use is the megapascal (MPa). Strain measures the deformation of an elastomer when a stress is applied to it. Here, we will only consider homogeneous strain or equal deformation from all parts of the part resulting from a uniform compound. It can be represented as

$$\varepsilon = \frac{l - l_0}{l_0} \tag{2.8}$$

Here, l_0 is the original length of the elastomer, and l is the length of the elastomer under load. There are no units for strain. To get an idea of the elasticity of an object, we discuss the modulus of elasticity, or "Young's modulus," also denoted by E. This value is the slope of the stress–strain curve and is a measure of the material's tendency to deform in an elastic manner. Its units are typically megapascals and it is represented by

$$\lambda = \frac{\sigma}{\varepsilon} \tag{2.9}$$

The Poisson ratio can be best described as the material's ability to contract in one direction when stretched in another. If we take one direction as axial and another as transverse, then

$$\nu = -\frac{\varepsilon_{trans}}{\varepsilon_{axial}} \tag{2.10}$$

Thus, as an elastomer is compressed in, say, the Z-direction (as in an isolator on a rubber grommet, engine mount, or transmission mount), the mount will deform in the X and Y directions. This value is nearly 0.5 for natural rubbers (typically used for mounts in automotive systems). For steel, Poisson ratios are around 0.3. The Poisson ratio has no units.

Other important parameters are the shear modulus or modulus or rigidity (G), which is the mount's (or other elastomeric components') ability to resist shear when forces are applied in opposing directions. For instance, there could be engine torque applying a load in a particular direction upon hard acceleration, while simultaneously a road impact force could be applied to the frame in another direction. This quantity is represented by the shear stress (τ) over shear strain (ε). Finally, the bulk modulus (K) plays a role in these types of components. The bulk modulus describes how a component elastomer will behave under pressure in three dimensions. Volume is considered here and typical units are in gigapascals (GPa). Equation 2.11 describes the bulk modulus mathematically, and Figure 2.7 shows the value graphically.

$$K = -V \frac{\partial p}{\partial V} \tag{2.11}$$

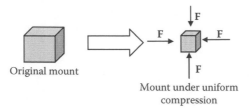

FIGURE 2.7   Rubber mount under compression.

V is volume, and the partial derivative of pressure with respect to volume accounts for the volume change (Figure 2.7).

As for the traditional role, the measurement of failed components and inspection of incoming components and raw materials (not original equipment manufacturer, OEM) has been among the duties that chemists have performed. We will discuss the design aspects of elastomers in Chapter 8.

2.12   PLASTIC AND ELASTOMER ANALYSIS

An Instron tensile machine can be used to measure stress and strain in the automotive lab. However, incoming inspection at a supplier is generally where this type of analysis takes place. More often, an automotive chemist at an OEM will be required to analyze a sample for failure analysis of that particular part. A thermogravimetric analyzer (TGA) will measure the change in mass of an elastomer in an inert (nitrogen) atmosphere as the temperature is waived. These instruments are very precise and will give information such as degradation temperature, filler content solvent residue, and absorbed moisture content.

Another useful tool that the automotive chemist will use is the differential scanning calorimeter. The component will be measured against a reference as heat is applied to both samples. A constant temperature will be maintained and the difference in temperature (ΔT) required to raise the temperature of the sample in relation to the reference sample is measured. Because the reference sample is not undergoing phase transitions, the ΔT of the measured sample will show an endothermic (melting point) or exothermic (crystallization) peak as heat is applied [7–9].

2.13   DSC GRAPHS

Figure 2.8 shows a sample differential scanning calorimetry (DSC) graph. The first peak along the curve shows the T_g. At the glass transition temperature, a polymer will exhibit a change in heat capacity. Below the T_g, polymers will be in a glassy state [10]. As heat is continually applied, molecules will have enough freedom of motion to align themselves into a crystalline structure. This will constitute a phase transition that will be exothermic and the crystallization temperature T_c can be measured. The area underneath the curve represents the enthalpy of the transition ΔH. ΔH is the area under the curve times the calorimetric constant for the machine, K:

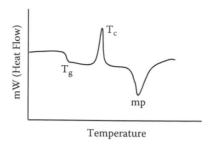

FIGURE 2.8   Example DSC trace.

$$\Delta H = KA \tag{2.12}$$

The knowledge of the enthalpy of peaks after T_g gives the automotive chemist knowledge of the cross-linking process. This is an important property for isolators because the impedance of vibrational inputs is dependent on either the mass of a system or the ability of an isolator or mount to absorb vibration. Generally, if a component's hardness is influenced more by cross-linking than by addition of filler (i.e., carbon black, which can be measured by TGA) the ability to absorb NVH inputs is better.

2.14   STRESS–STRAIN RELATIONSHIPS

Figure 2.9 shows an overlay of the stress–strain curve of natural rubber and an impact-modified nylon 6,6. At the top of the nylon curve is the yield point or elastic limit of the nylon. At this point, the strain is no longer recoverable and the sample exhibits plastic deformation. Of course, below this point, the material obeys Hook's law ($F = -kx$), where k is the spring constant, x is the distance traveled when force is applied, and F is that applied force.

In the design of gaskets and other rubber components, the chemist is often asked to provide material recommendations or values to the CAE analyst for input into a

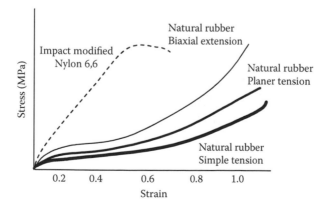

FIGURE 2.9   Stress–strain curves of elastomers versus thermoplastic.

model. The analyst generally uses a program such as ABAQUS or MSC.MARC. These programs work very well when designing in plastic, as we will see in later chapters. However, for rubber compounds, the linear shape of the graph that follows Hook's law for plastic no longer applies. Miller [11] recommends an experimental approach of (a) establishing large strain nonlinear equilibrium material properties in the range of interest, and (b) determining the dynamic properties of superimposed linear vibrations in the frequency, mean strain, and dynamic amplitude in the range of interest.

Analysis of small-amplitude vibrations has been reported by Morman as applicable to rubber support mounts, automotive door seals, and vibration isolation gaskets for frequencies between 0.01 and 20,000 Hz [12]. According to Miller, hyperelastic material models require three sets of stress strain curves. These are the three curves that need to be input (shown in Figure 2.9). They are biaxial extension, planar extension, and simple tension. Data must also be gathered on a biaxial machine [13]. A biaxial test will be performed along multiple axes of the rubber sample without testing to failure. This will result in a more accurate model along the usable range of the material. It will also discount the effects of friction.

2.15   BOND STIFFNESS VERSUS MODULUS

Elastomers are long chained polymers that are bent back upon themselves many times. They have moduli of less than 1 GPa. The segments of the molecules will freely slide back across each other and are strengthened by fillers and the occasional cross-link. Table 2.7 shows bond type versus stiffness and modulus. With increasing temperature, entropy or disorder increases. Thus, stretching these types of parts decreases their entropy because molecules are aligned to become more crystalline.

TABLE 2.7
Bond Type versus Stiffness and Modulus

Bond Type	Bond Stiffness (N/m)	Young's Modulus (GPa)	Example
Covalent	50–80	200–1000	C–C
Metallic	15–75	60–300	Metals
Ionic	8–24	32–96	NaCl
Hydrogen	6–3	2–12	H_2O
van der Walls	0.5–1	1–4	Waxes

REFERENCES

1. Skoog, D., and D. West. 1980. *Principles of instrumental analysis,* 2, 303, 336–337, 674–675. Philadelphia: Saunders College.
2. Smith, S. B., Jr., and G. M. Hieftje. 1983. A new background-correction method for atomic absorption spectrometry. *Applied Spectroscopy* 37 (5): 419–424.
3. Watanabe, S. 2007. Characteristic properties of water-soluble cutting fluid additives for iron materials. *Journal of Oleo Science* 56:385–386.
4. Watanabe, S. 1998. Perpetration and characteristic properties of water-based cutting fluids additives. *Recent Developments in Oil Chemistry* (Transworld Research Network) 2:145–187.
5. Poole, C., and S. Schuette. 1985. *Contemporary practice of chromatography,* 375. New York: Elsevier.
6. Martin, A. J. P., and R. L. M. Synge. 1941. A new form of chromatogram employing two liquid phases. *Biochemical Journal* 35:1358.
7. Dean, J. A. 1995. *The analytical chemistry handbook,* 15.1–15.5. New York: McGraw–Hill.
8. Pungor, E. 1995. *A practical guide to instrumental analysis,* 181–191. Boca Raton, FL: CRC Press.
9. Skoog, D. A., F. J. Holler, and T. Nieman. 1998. *Principles of instrumental analysis,* 5th ed., 905–908. New York: Brooks Cole.
10. Varshneya, A. 1994. *Fundamentals of inorganic glasses.* Boston: Academic Press.
11. Miller, K. 2000. *Measuring dynamic properties of elastomers for analysis.* DynamicRev1. Ann Arbor, MI: Axel Products Inc. www.axelproducts.com
12. Morman, K. N., Jr. 1996. Analytical prediction of sound transmission through automotive door seal systems. Presented at *Third Meeting of the Acoustical Society of America and the Acoustical Society of Japan,* Honolulu, Hawaii, December 2–6.
13. Morman, K. N., Jr., B. G. Kao, and J. C. Nagtegaal. 1981. Finite element analysis of visoelastic elastomeric structures vibrating about nonlinear statically stress configurations. SAE paper 811309.

3 Component Materials in Automobiles

3.1 INTRODUCTION

In this chapter we will discuss the basic families of materials used in today's auto-mobiles. We will concentrate on polymers, although the chemist's role deals with all materials used in a vehicle. The use of polymers has steadily increased ever since their introduction into a vehicle's structure. Exterior as well as interior parts and trims made from polymers are increasingly replacing traditional steel or aluminum parts [1]. The drive for mass and cost savings is enhanced by the material's ability to be custom-ized and to decrease material cost. Coatings are applied to interior parts as well as to exterior parts to enhance appearance and provide corrosion resistance. Stabilization is required of many of the polymeric components (this topic is addressed in Chapter 4).

3.2 POLYMER MARKET PENETRATION

One of the best methods to fulfill each of the four competitive imperatives is to use polymer materials in the vehicle. Polymers can be considered as catalysts [1] provid-ing for

- technical innovations (bumpers and fascias to meet impact regulations);
- economic innovations (less expensive than many other materials);
- aesthetic innovations (increased design freedom); and
- environmental innovations (recycling ability and reduction of fuel consumption).

These innovations have led to an increase in the use of polymers in automobiles from about 30 kg per vehicle in the 1970s to more than 125 kg today for the North American market [2]. This is shown in Table 3.1. The progression of polymer use in the European market by country is shown in Table 3.2 [3].

We can see that a similar progression for Europe is shown from 1970 to 1990. The distribution of materials in an average size vehicle of 1,300 kg (2,866 lb) is shown in Table 3.4. The polymer percentage has increased to 114 kg today versus the 30 kg or so utilized in the past. The increased polymer usage represents a substantial mass savings. Body structures represent the largest growth opportunity in the future [1]. This includes fascias, wheel frames, body panels, and entire roof modules.

Areas in a vehicle where polymers are used are shown in Table 3.3. The chemist and design engineer in the future will take the opportunity to make improvements

TABLE 3.1
Polymer Usage in North American Vehicles

Year	Kilograms/Vehicle
1970	31.75
1988	68.04
1999	116.57
2000	115.67
2005	126.55
2010	139.25

Source: Bechtold, K. *Material Testing Product News,* 36, 77, 2006. With permission.

TABLE 3.2
European Polymer Usage in Vehicles by Year

Country	1970 (kg)	1980 (kg)	1990 (kg)
France	39	71	98
Germany	56	80	104
Italy	39	79	98
Western Europe	47	77	100

Source: Bechtold, K. *Material Testing Product News,* 36, 77, 2006. With permission.

TABLE 3.3
Applications for Polymers in Vehicles

Automobile Exterior	Automobile Interior	Under the Hood
Bumpers	Instrument panels	Drive belts
Front ends	Doors pillars, side trims	Air intake manifolds
Grills	Consoles	Gaskets and membranes
Mirrors	Seats	Valve covers
Handles and locks	Roof liners	Noise dampening
Sliding roofs	Steering wheels	Fans, shrouds, and tension pulleys
Body components	Heating/air conditioning	Fuel supply components
Wheel covers	Cable tree/lighting	Air supply components
Headlamps	Floor lining	Oil supply components
Taillights	Pedals	Water supply components
Windshield wipers	Engineering components	HVAC components
Glazing		Transmission seals
Decorative trim		Engine covers
Shock absorption		Electrical harnesses
Underbody protection		Dust caps and sleeves
Seals		

Source: Bechtold, K. *Material Testing Product News,* 36, 77, 2006. With permission.

TABLE 3.4
Material Mass in Vehicles

Material	Mass (kg)	Percentage
Steel	807.8	60.66
Iron	156.5	11.75
Aluminum	116.2	8.73
Copper and brass	20.9	1.57
Zinc	5	0.37
Polymers	114.8	8.62
Rubber	66	4.96
Glass	44.7	3.35
Total	1331.9	100.00

Source: Bechtold, K. *Material Testing Product News,* 36, 77, 2006. With permission.

in these areas for design, cost, and other innovations to make vehicles lighter and more efficient.

In Table 3.4, looking at a breakdown of the total vehicle mass and the materials used within it, we can see the expected large amount of steel at 800 kg, followed by iron at 150 kg, and aluminum at 116 kg. If we consider a vehicle's life to be over 100,000 miles, then the associated weight reduction from using 8% polymers versus 4% would translate into a savings of 750 L of fuel [1]. Further development of lighter weight materials (not only polymers, but also aluminum and magnesium) is a very high priority with auto original equipment manufacturers (OEMs).

Table 3.5 presents the various types of polymer materials used in a vehicle, their quantities, and their approximate mass. At 20 kg, interior trim represents the highest amount of polymer usage, followed closely by the dashboard and seats.

In Table 3.6 we examine some of these automotive materials, their properties, some usages, and the structure of the molecule. These materials are typically chosen by the material engineer, who works closely with the design engineer from a decision tree. In the decision tree, the usage environment is considered, including normal use temperature, the various loads on the part, appearance criteria, maximum use temperature, and tolerance requirements. This decision tree is usually in the form of a flow diagram that will help the engineer and chemist make the material decision.

3.3   METHODS OF PRODUCTION AND PRODUCTION DEMAND

Polyethylene (PE). High-density polyethylene (HDPE) was introduced in the 1950s. Production is typically from metal oxide catalysis. HDPE is 90% crystalline with a T_m of 135°C. In 2001, the production for HDPE was 12,320 million lb; in 2002, it was 13,600 million lb and, in 2006, 16,220 million lb. Historical growth has been on the order of 3.0–4.5% [4]. Recent price data ranged from a high of $0.54 to a low of $0.315 per pound.

TABLE 3.5

Type and Mass of Materials per Vehicle Part

Part	Main Plastic Type	Mass in Vehicle
Bumpers	PP, ABS, PC	10
Seats	PUR, PP, PVC, ABS, PA	13
Dashboard	PP, ABS, PA, PC, PE	15
Fuel systems	PE, POM, PA, PP	7
Body	PP, PPE, UP	6
Under-hood components	PA, PP, PBT	9
Interior trim	PP, ABS, PET, POM, PVC	20
Electrical components	PP, PE, PBT, PA, PVC	7
Exterior trim	ABS, PA, PBT, ASA, PP	4
Lighting	PP, PC, ABS, PMMA, UP	5
Upholstery	PVC, PUR, PP, PE	8
Reservoirs	PP, PE, PA	1
Total		105

Source: Bechtold, K. *Material Testing Product News,* 36, 77, 2006. With permission.

Polypropylene (PP). Polypropylene is manufactured by the Ziegler–Natta process. PP has a T_m of 170°C. Production demand for 2001 was 36,718 million lb; in 2002, it was 41,143 million lb and, in 2006, 50,000 million lb [4]. Historical growth has been 6.3% per year (1997–2002). Pricing has hit a high of $27.50 and a low of $12.00 per pound for polymer grade since 1997 [4].

Acrylonitrile-butadiene-styrene (ABS). ABS is manufactured through graft polymerization of acrylonitrile and styrene on a polybutadiene elastomer. The demand for this polymer was slightly over 1.40 billion lb in 2005 [4]. The price has ranged from a high of $1.05 to a low of $0.69 per pound [4].

Polyvinylchloride (PVC). Polyvinylchloride is manufactured by suspension or bulk polymerization methods. The demand in 2006 was 15.6 million lb, up from 13.2 million lb in 2001. Polyvinylchloride demand is expected to grow by approximately 2% per year over [4] the next 10 years. Price ranged from a low of $0.175 to a high of $0.41 per pound from 1997 to 2002 [4].

Polyethylene terephthalate (PET). PET is made by direct reaction of a diacid and diol or by ester interchange of a diester and a diol. PET is the most commercially useful polyester. The demand in 2006 was 11,500 million lb, up from 9,450 million lbs in 2001 [4]. The price of polyethylene terephthalate ranged from $0.43 to $0.71 per pound from 1997 to 2002 [4].

Polymethyl methacrylate (PMM). Polymethyl methacrylate is manufactured by the acetone cyanohydrin process, which is the only process used in the United States for the manufacturing of polymethyl methacrylate [5]. The demand in 2007 for this polymer was 1,485 million lb. In 2002, the demand was 1,285 million lb [6]. Price ranged from around $0.75 to $0.41 per pound from 1998 to 2003.

TABLE 3.6
Typical Polymers and Their Use in Vehicles

Polymer	Acronym	Properties	Application Examples	Structure
Polypropylene	PP	Low cost, good chemical resistance	Bumpers, wheel housings, air filter housings, guide channels, side panels	
Polyurethane	PUR	Good NVH dampening, good elasticity, low heat conductivity	Seat upholstery, jounce bumpers, dash and roof padding, exterior elements	
Acrylonitrile butadiene styrene	ABS	Electroplatable, dimensionally stable, good structure	Interior paneling, wheel panels, grills, fascias	
Polyamide	PA	Temperature stable, low gas permeability, heat resistant, structurally solid, dimensionally stable, age resistant, rigid, extremely versatile	Housings, structural applications, engine covers, ducting, radiator tanks, knobs, fasteners, wheel panels, plugs, suction elbows, covers	
Polyvinylchloride	PVC	Weather resistant, low cost, nonflammables	Underbody protection, protective bordering, cable insulation, interior paneling	
Polyethylene	PE	Low cost, ageing resistant, chemically resistant, good structure	Fuel tanks, windshield fluid containers	

—continued

TABLE 3.6 (continued)
Typical Polymers and Their Use in Vehicles

Polymer	Acronym	Properties	Application Examples	Structure
Polyoxymethylene	POM	Chemical resistance, abrasion and impact resistant, low tendency to creep, thermally stable	Clips, connectors, bearing components	
Polymethyl methacrylate	PMMA	Transparent, scratch and UV resistant, stress-cracking resistant	Headlight lenses for blinker and rear lamps	
Polycarbonate	PC	Impact and UV resistant, transparent	Headlight lenses, bumper coverings, exterior auto body parts	
Polyethylene terephthalate	PET	Good tensile strength and rigidity, effective barrier for gases	Textiles, coverings	
Polybutylene terephthalate	PBT	Rigid, heat resistant, good electrical insulating behavior, dimensional accuracy	Electronic housings, bumper coverings, exterior auto body parts, plugs	

Polycarbonate (PC). Polycarbonate is similar to the chemical behavior of polyesters. It is produced from a reaction of bisphenol A [2,2-bis (4-hydroxyphenyl)-propane] and phosgene. The price has ranged from $1.29 per pound to $1.36 per pound [6]. Demand for polyurethane was 1,440 million lb in 1996 and is expected to grow at a rate of 5.2% annually [7].

Polyoxymethylene (POM). Polyoxymethylene can be made from a condensation reaction of polyformaldehyde (DuPont method). Demand for polyacetal has been around 4,600 tons per year [8]. The price for polyacetal resin is $1.44 per pound [9].

Polyurethane (PU). Automotive polyurethanes can be measured from a reaction of isocyanates and alcohol. Many automotive foams use toluene diisocyanate (TDI). The demand for polyurethane is projected to reach $7.4 billion in resin cost in 2009. This represents an increase of 3.2% from 2008. The total demand is expected to be 2.8 billion lb in 2009, up by 4.5% [10]. The price varies greatly, depending on the type of polyurethane used.

Polyamide (PA). Depending on the polyamide (nylon 6 or nylon 6,6), the manufacturing process is a ring opening reaction of a caprolactam for nylon 6 or step growth polymerization in the case of nylon 6,6. Demand for the polymer was 3,050 million lb in 2001 and 2,006 million lb in 2006 [5]. Growth of this polymer overall is negative, and in the automobile industry certain applications are increasingly being replaced by polypropylene. These applications include such things as fan shrouds for electric cooling fans. From 1997 to 2003, the price ranged from $1.27 to $1.47 per pound.

3.4   ZIEGLER–NATTA

Karl Ziegler and Giulio Natta discovered a catalyst in the 1950s that gave stereo regularity for polymerized alpha olefins and dienes. Figure 3.1 shows the isotactic polypropylene given by the Zieglar–Natta method. Figure 3.2 shows the alternating syndiotactic structure and random atactic structure given if polypropylene is produced by a different method.

Ziegler–Natta catalysts are typically based on titanium compounds, such as $TiCl_4$, along with organometallic aluminum compounds of the general formula R_3Al, such as triethylaluminum $(C_2H_5)_3Al$. Ziegler and Natta were awarded the Nobel Prize in chemistry in 1963 for this work. The polymers that Natta produced can be isotactic, syndiotactic, or atactic, depending on the orientation of the alkyl groups in the polymer chain. Chiral centers in syndiotactic polymers alternate their stereochemistry,

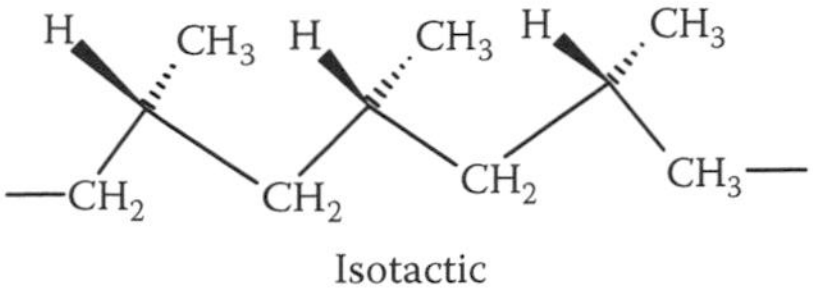

Isotactic

FIGURE 3.1   Isotactic polypropylene.

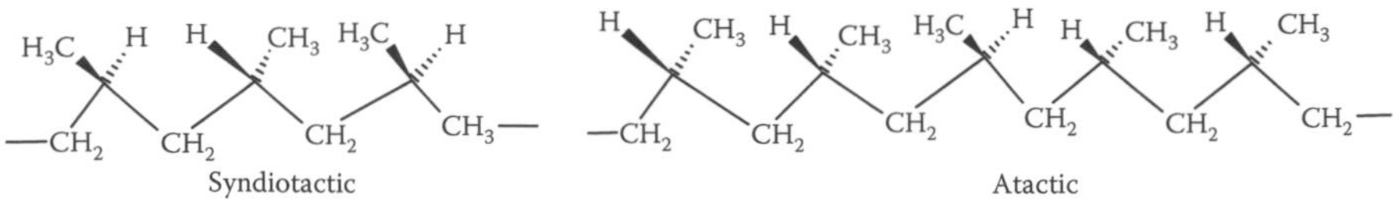

FIGURE 3.2 Syndiotactic and atactic polypropylene.

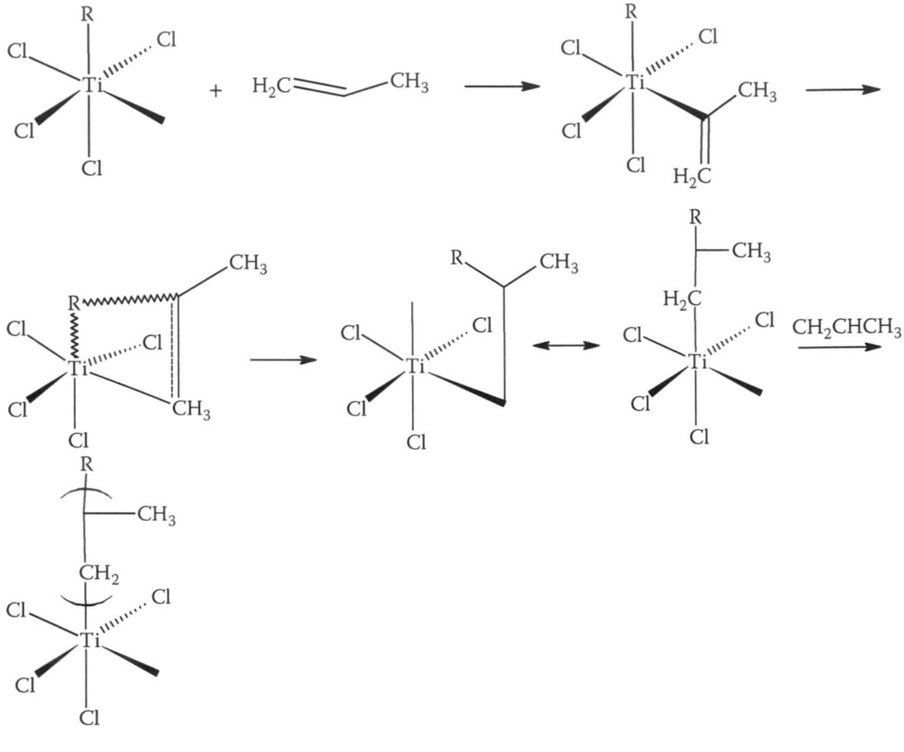

FIGURE 3.3 TiCl$_4$ catalyst mechanism.

but atactic polymers do not have a regular repeating pattern. The mechanism for the reaction is partially understood [5]. The TiCl$_4$ derived catalyst will convert polypropylene to isotactic polymers; other systems, such as VCl$_4$, are used to convert polypropylene to syndiotactic polymers [11].

The titanium salt and the organometallic will react to give a penta-coordinated titanium complex with a sixth empty site of the octahedral type [5]. Looking at Figure 3.3, we can see the sixth site of the octahedral complex open. An alkene monomer is complexed with the titanium and inserts itself between the titanium and alkyl group; this would leave a new empty site for the process to repeat [5]. Polypropylene, polyvinylalcohol, and poly alpha olefins can be produced by Ziefler–Natta. Five different polybutadienes can be made from Ziegler–Natta, either the 1,2 or 1,4 configuration. The 1,4 configuration has a double bond that can be atactic, syndiotactic, or isotactic [5].

The sixth site of the Ti centers comprises defects, where some Ti centers lack their full complement of chloride ligands. The alkene molecule will bind at these vacancies. In ways that are still not fully clear, the alkene converts to an alkyl ligand group. The stereospecific nature of the alkane produced is facilitated by restriction of the coordination sphere of the titanium atom [12].

A large number of different Ziegler–Natta catalysts were developed during the past 30 to 40 years. Metallocene catalysts have been developed as well—partly to help understand the mechanism of Ziegler–Natta. Metallocenes are compounds with the general formula $(C_5R_5)_2M$, which has two cyclopentadienyl anions bound to a metal center. Group 4 metallocene derivatives $[CpZrCH_3]^+$ catalyze polyolefins. By definition, these compounds contain a transition metal and two cyclopentadienyl anions that are coplanar with equal bond lengths and strengths. One of the features of some metallocenes is their high thermal stability. These metallocenes can have either an eclipsed or a staggered structure.

3.5 METAL OXIDE INITIATION

High-density polyethylenes are formed from metal oxide initiation. These poly-ethylenes are stiffer, with much less branching. HDPE is used more in automo-tive applications due to this property. Phillips Petroleum uses a chromic oxide catalyst and Standard Oil of Indiana has developed a molybdenum oxide cata-lyst [5]. These catalysts are not flammable, which presents an advantage over Ziegler–Natta.

3.6 OTHER METHODS OF PRODUCTION

Polymer processing can be of several types, including free radical, cationic, anionic, metal complex, or metal oxide catalyzed, as mentioned earlier [5]. Polymers can be made by bulk polymerization, solution polymerization, suspension polymeriza-tion, or emulsion polymerization techniques [5]. The automotive chemist or design engineer working for an OEM should be aware of these various manufacturing pro-cesses, which polymers are made by which process, and what characteristics can be expected from the type of process.

3.7 CHAIN GROWTH POLYMERIZATION

Chain growth polymerization has the characteristic of having an intermediate within the process that cannot be isolated [5]. The intermediate can be a metal complex, a free radical, or an ion. These intermediates are transient to the process. The terms *vinyl*, *olefin*, and *addition* polymerization have been associated with this process [13]. Monomer units add to a chain very rapidly once it has been initiated. Initiation is the creation of an active center such as a free radical or carbanion [13]. An example is the thermal decomposition of benzoyl peroxide shown in Figure 3.4. To propagate the chain, an additional monomer is added at a very rapid rate as monomer concen-tration is reduced. Figure 3.5 shows the propagation of polystyrene.

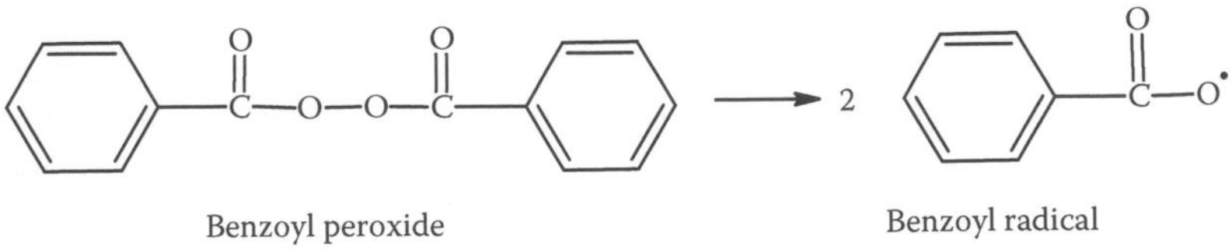

FIGURE 3.4   Thermal decomposition of benzoyl peroxide.

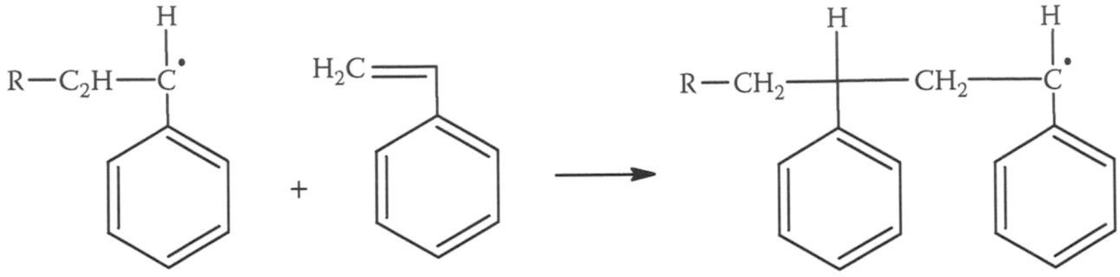

FIGURE 3.5   Polystyrene propagation.

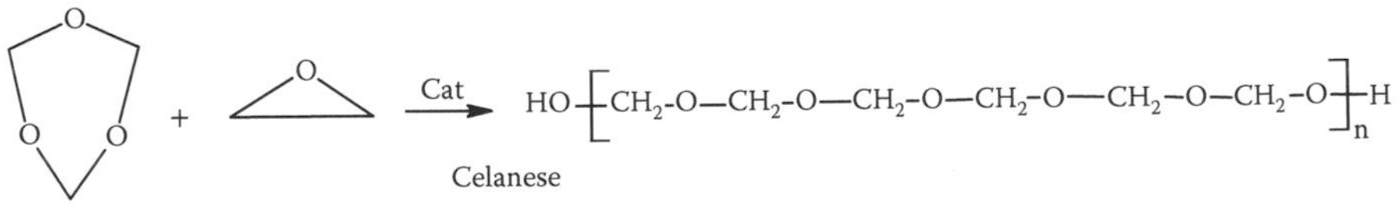

FIGURE 3.6   Celanese production method of POM.

Chain growth polymerizations very often contain a double bond; however, cyclic ethers will polymerize in this manner [5]. POM (polyoxymethylene) made by the Celanese method shown in Figure 3.6 is an example of a cyclic ether with this method. The Celanese route for the production of polyacetal yields a more stable copolymer product via the reaction of trioxane, a cyclic trimer of formaldehyde, and a cyclic ether (e.g., ethylene oxide or 1,3 dioxalane).

The DuPont route for polyacetal (shown in Figure 3.7) yields a homopolymer through the condensation reaction of polyformaldehyde and acetic acid (or acetic anhydride). Termination can be achieved by coupling or disproportionation [13]. Termination can be thought of as the disappearance of an "active" center [13]. Figures 3.8 and 3.9 show disproportionation and coupling, respectively. The two active centers will combine in the case of coupling, while termination via disproportionation leaves a double bond on one of the chain ends as the reaction will simply end.

The initiation, propagation, and termination shorthand notations are shown in the following equations [13].

Initiation:

$$I \xrightarrow{\;ki\;} 2R\bullet \qquad\qquad\qquad (3.1)$$

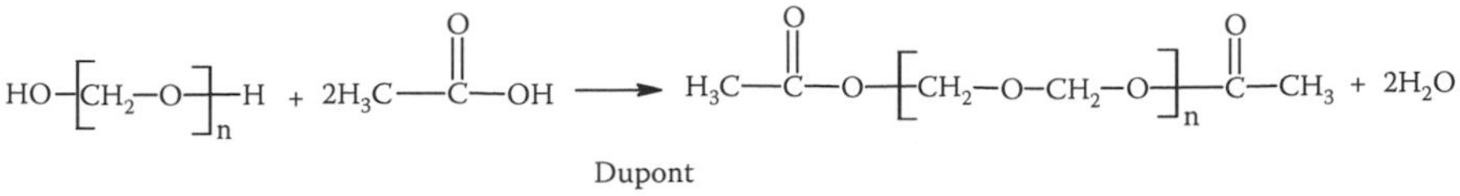

FIGURE 3.7 DuPont production method of polyacetal.

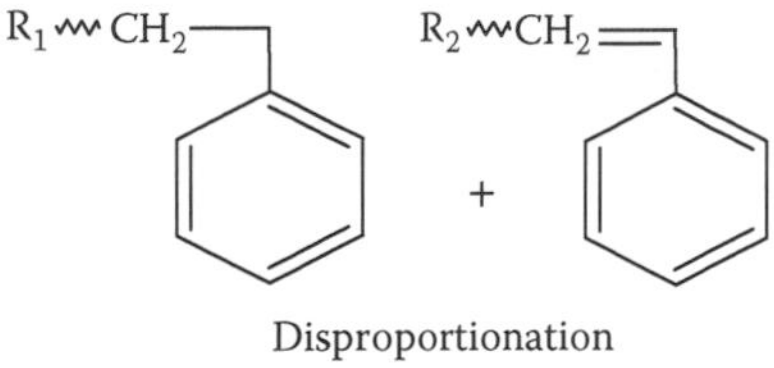

FIGURE 3.8 Termination by disproportionation.

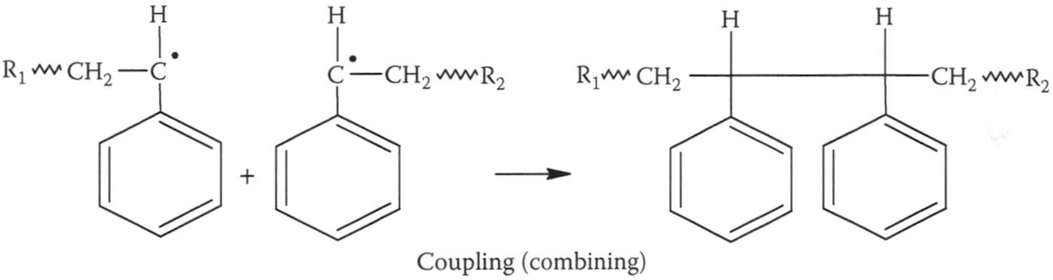

FIGURE 3.9 Termination by coupling.

$$R\bullet + M \xrightarrow{k2} RM\bullet \qquad (3.2)$$

Propagation:

$$RM\bullet + M \xrightarrow{kp} RM_2\bullet \qquad (3.3)$$

Termination:

$$RM_n\bullet + RM_m\bullet \xrightarrow{kt} R_2M_{n+m} \qquad (3.4)$$

The rate constant is k, M is the monomer, and R is the initiating radical. A polymer's reactivity depends on the last unit added to the growing chain rather than on the length of the chain or any of the other units on the chain [13]. Propagation is the dominant step and therefore the only effective step that consumes the monomer. The rate of polymerization (R_p) is

$$R_p = -\frac{d[M]}{dt} = k_p[M\bullet][M] \tag{3.5}$$

3.8 STEP GROWTH POLYMERIZATION

Nylon 6,6 and nylon 6 are hugely important in the automotive industry. Nylon's strength and versatility allow for use under the hood as well as in the interior. When designing, a chemist or design engineer's first choice of materials begins with nylon. Nylon is produced by step growth polymerization. Step growth can be categorized by reactions of molecules with functional groups. This process can be stopped and low molecular weight oligomers can be obtained [5]. Monomer concentration will not decrease at the rate that it will in chain growth polymerization, but it will decrease at a fast rate early in the reaction due to the formation of oligomers [13]. After the reaction begins, there is a distribution of oligomers that are reacting at a slower rate. As we can see from the production of nylon 6,6 (Figure 3.10), a molecule of water is produced when the polymer reacts. For this reason, the term *condensation polymerization* is often used. When nylon 6 is produced by step growth, water is not produced (Figure 3.11).

3.9 IONIC POLYMERIZATION

The electron-withdrawing nature of the constituent on a vinyl monomer (CH_2=CHX) will affect the polymerization of that monomer. If the constituent is electron donating, then a cationic initiation mechanism is favored. If the constituent is electron withdrawing, then an anionic mechanism is favored.

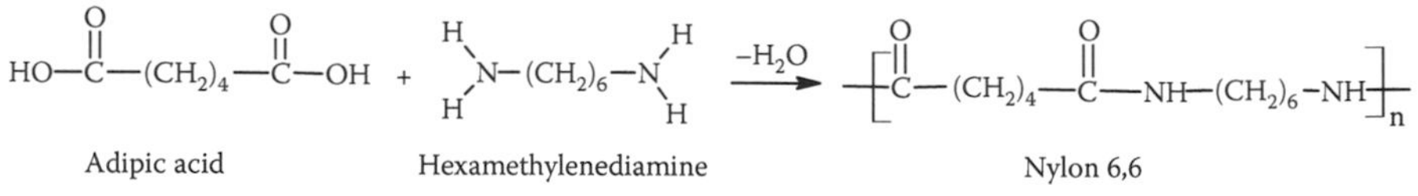

Adipic acid   Hexamethylenediamine   Nylon 6,6

FIGURE 3.10   Nylon 6,6 production.

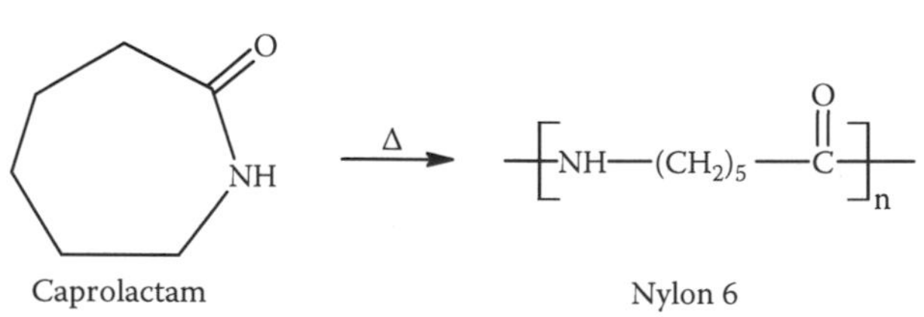

Caprolactam   Nylon 6

FIGURE 3.11   Nylon 6 production by ring opening caprolactam.

TABLE 3.7
Types of Chain Polymerizations for Various Monomers

Cationic Only	Free Radical Only	Anionic Only	Cationic or Free Radical	Free Radical or Anionic
Isobutylene	Halogenated	Vinylidene	N-vinylcarbazole	Acrylic and
Alkyl vinyl ethers	vinyls	cyanide	N-vinylpyrrolidone	methacrylic esters
Derivatives of	Vinyl esters	Related cyano		Vinylidene esters
α-methylstyrene		derivatives		Derivatives of
		Nitroethylenes		acrylonitrile

Schildknecht classified the type of chain polymerization suitable for the various monomers shown in Table 3.7 [13]. In the various ionic methods of polymerization, the monomer must fit between the growing chain end and an ion complex [13]. Cationic polymerizations proceed very rapidly, with the lifetimes of growing chains less than 1 s in the case of isobutylene. Stereoregularity is obtained as monomers are fit between chain and counter ions when polymerized.

When lithium alkyls are used in anionic polymerizations, they tend to give *cis* products (*cis*-polybutadiene and *cis*-polyisoprene). There is also no termination step with these polymerizations. The rate of polymerization depends on the amounts of initiator and monomer present [13].

REFERENCES

1. Bechtold, K., and H. Clariant. 2006. "String Combinations," *Material Testing Product and Technology News* 36:1, 3.
2. Automotive Plastics Report 1999–2000. Market Search, Inc. http://www.plasticstoday.com
3. Mavel report. http://www.mavel.com
4. http://www.the-innovation-group.com/
5. Chenier, P. J. 1992. *Survey of industrial chemistry,* 254, 272–273, 280–281. New York: VCH.
6. http://www.tgdaily.com/content/view/25825/122/
7. http://www.theiapdmagazine.com/pdf/magazine-archives/31.pdf
8. http://www.dsir.gov.in/reports/techreps/tsr162.pdf
9. http://www.ides.com/resinprice/resinpricingreport.asp
10. http://www.the-infoshop.com/study/fd36644-polyurethane.html
11. Hill, A. F. 2002. *Organotransition metal chemistry,* 136–139. New York: Wiley-InterScience.
12. Bochmann, M. 1994. *Organometallics 2: Complexes with transition metal–carbon π-bonds,* 57–58. New York: Oxford University Press.
13. Rodriguez, F. 1989. *Principles of polymer systems,* 3rd ed., 77–79, 86, 88. New York: Hemisphere Publishing.

4 Design Concerns and Imperatives

4.1 INTRODUCTION

In Chapter 1 we discussed the four competitive imperatives that the design chemist must consider as a vehicle is developed: performance, mass, cost, and environmental considerations. All four must be considered as a vehicle is developed.

Of all the aspects of vehicle production, including design, development, and manufacturing, the greatest opportunity to affect these imperatives is in the design phase of vehicle development. During the past 15 years, consumers have leaned toward larger utility vehicles while there has been a simultaneous demand to reduce emissions and save gas. During this latest crisis, automakers have been working to develop advanced technologies to address the important concerns around environmental issues and energy. As we are tasked to make improvements in performance, mass, and cost, we must complete this task up front in the design phase. Fortunately, we have tools such as finite element analysis and other predictive software to optimize designs before they are manufactured.

4.2 HISTORY OF AUTOMOTIVE DESIGN

The history of automobile design in the United States was affected to a great extent by Alfred P. Sloan, Jr., president of General Motors in the 1920s. Sloan implemented annual model year design changes in order to convince car owners that they needed to purchase new vehicles each year. This was due in part to the fact that the automobile market had become saturated in 1924. With this "dynamic obsolescence" strategy, the face of design was changed. The process of design and redesign is expensive and this marketing strategy could not be matched by many of the smaller car manufacturers. Ford Motor Company sought to keep vehicles simple while maintaining design integrity. This philosophy was later adopted by Ferdinand Porsche.

Another effect of GM's strategy was to draw design away from aerodynamic concepts that produce a single shape, such as a Porsche 911. Adopting the Ford or Porsche strategy would have been contrary to the new GM direction and affected sales of the new vehicle styles. Body-on-frame design was implemented versus flexible monocoque designs in order to enable more frequent changeovers. The strategy did pay off when, in 1931, GM outsold Ford in vehicle sales [1]. Styling was forever changed and the union of engineering and design has been difficult ever since. Cosmetic changes take place every few years on a particular vehicle platform.

Until recently, vehicle sales were mainly based on consumer expectations rather than engineering advances.

4.3   AUTOMOTIVE DESIGN DEVELOPMENT

Automotive design involves development of the appearance of a vehicle and the ergonomic integration of the components. The design and development of a vehicle is done by a large team from many different disciplines. Designers are typically degreed in industrial or transportation design or have an art background. The task of the design team is typically split into exterior design, interior design, and trim. The role of the chemist, of course, revolves around the materials involved with these designs. Design focuses on all aspects of the vehicle rather than on simply the outer shape of automobile parts. Power train applications, of course, are not involved in the process, but are chosen based on the performance requirements of the vehicle. The steps in the vehicle design process include:

- concept sketching;
- computer-aided design (CAD);
- computer modeling;
- drive-train engineering;
- scale model creation;
- prototype development; and
- manufacturing process design.

4.3.1   EXTERIOR DESIGN (STYLING)

The design staff responsible for the vehicle's exterior design will develop concept sketches. These can be done manually or with computer help. The proportions, shapes, and surfaces of the vehicle are created. The drawings are refined progressively until they are at the point at which the design staff feels that no further development is needed. Digital or computer models are developed along with clay models at this time. Clay models are used in the first design iterations. Clay occurs naturally and is composed of fine-grained minerals. The clay mineral group is composed of the following:

kalonite: $Al_2Si_2O_5(OH)_4$;
illite: $(K,H_3O)(AlMgFe)_2(Si,Al)_4O_{10}[(OH)_2,(H_2O)]$;
montmorillonite: $(NaCa)_{0.33}(Al,Mg)_2(Si_4O_{10})(OH)_2$ nH_2O;
talc: $Mg_3Si_4O_{10}(OH)_2$;
vermiculite: $(MgFe,Al)_3(Al,Si)_4O_{10}(OH)_2$ $4H_2O$; and
pyrophyllite: $Al_2Si_4O_{10}(OH)_2$.

Clay occurs naturally and has plasticity when water is added. These clay minerals are phyllosilicates, which have water trapped in the mineral structure by polar attraction [2]. The data from these solid models are then used to create a full-sized mock-up of the final design or body in white. The chemist's role is minimal at this point in the design process.

4.3.2 INTERIOR DESIGN

The design staff will develop the interior of the vehicle based on themes such as rectangular or round AC outlets, door trim panels, seats, ergonomic placements, and surface finishes on the IP (instrument panel). As with exteriors, sketches are made and then digital models are created, followed by clay models. Interior colors and material choices, including paints, fabrics, leather grains, carpets, and plastics, are made. The chemist's role here is to help with the choice of interior materials and finishes of the materials. Injection-molded plastics in interiors, such as instrument panels, trims, consoles, and sun visors, all have "styled finishes"—a surface treatment to the injection mold tool to give a desirable appearance or class "A" surface appearance. Class "A" appearance has curvature and tangency alignments to nearly perfect aesthetic reflection quality.

In addition, the color of the material must be matched to the theme of the car and the leather or fabric seats being utilized. Designers will often draw inspiration from other design disciplines, such as industrial design, fashion, or aeronautics. Research is conducted into global trends in design for projects two to three model years in the future. Trend boards are created from this research in order to keep track of design influences as they relate to the automobile industry. Themes and concepts are developed to refine vehicle models further.

4.3.2.1 Interior Design and Performance

A major role of the chemist in interior design is to utilize coatings and stabilizers to improve the appearance and life of the product and protect the part from physical and chemical stress. The automobile industry is setting higher standards to fulfill the stabilization requirements for vehicle interior coatings. Some of the materials utilized in interior applications were listed in Chapter 3.

Coatings are generally applied to protect the underlying polymer or other substrate from mechanical and chemical stress and environmental impact. For example, the dash is exposed to ultraviolet rays on a daily basis. These rays are detrimental to the underlying polymer, which must be enhanced to improve the life and appearance of the dash. In addition to protection, coatings hide the differences in gloss and shade that occur due to injection molding. These coatings provide a perceived "higher value" of the product.

Coatings have to be protective and resistant to oxygen, humidity, staining, and weather. To protect against these effects delaminating the substrate, material stabilizers are added. There are three general categories of photostabilizers, each functioning by a different chemical mechanism [3]:

UV absorbers;
radical traps; and
quenchers.

TABLE 4.1

Activation Spectra Maxima for Some Polymers

Polymer	Activation max λ
Polyesters	325 nm
Polystyrene	318 nm
Polyethylene	300 nm
Polypropylene (unstabilized)	310 nm
Poly (vinyl chloride)	320 nm
Poly (vinyl acetate)	280 nm
Polycarbonate	295 nm
Polymethyl methacrylate	290–315 nm
Polyoxymethylene	300–320 nm
Cellulose acetate butyrate	295–298 nm

Source: Hawkins, W. L. *Polymer Degradiation and Stabilization*, Springer-Verlag, 1984. With permission.

Ultraviolet absorbers are the most widely utilized of the photostabilizers. Most common UV absorbers are low molecular weight derivatives of *o*-hydroxybenzophenone, *o*-hydroxybenzotriazole, or *o*-hydroxyphenyl salicylate [3]. Most polymers are sensitive to UV radiation between 300 and 360 nm. They absorb UV radiation through their structure; structural irregularities or impurities also have an effect [3]. Table 4.1 shows some activation spectra for some typical polymers [3]. Individual polymers absorb UV radiation within specific wave length regions, exhibiting activation spectra maxima at the wavelength where each is most vulnerable to photo-oxidation [4].

If a plot of UV absorbers versus wavelength is made, the resulting curves would give a guide to choosing the best performing ultraviolet absorber by choosing the one with the highest absorption peak in the region of interest. Hydroxybenzophenones are believed to function by converting UV energy into vibrational energy within a hydrogen bond [3] (Figure 4.1).

The mechanism for this conversion is the ability of these compounds to form a six-membered ring that contains a hydrogen bond [3]. *o*-Hydroxyphenylbenzotriazoles (Figure 4.2) and *o*-hydroxyphenyl salicylates (Figure 4.3) dissipate absorbed energy by this mechanism as well.

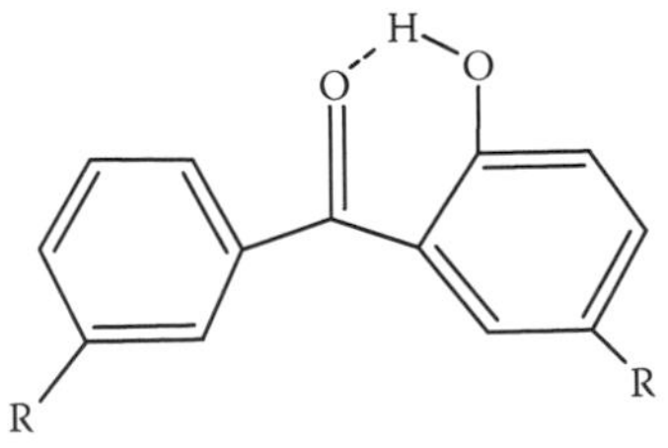

Hydroxybenzophenones

FIGURE 4.1 Structure of hydroxybenzophenones.

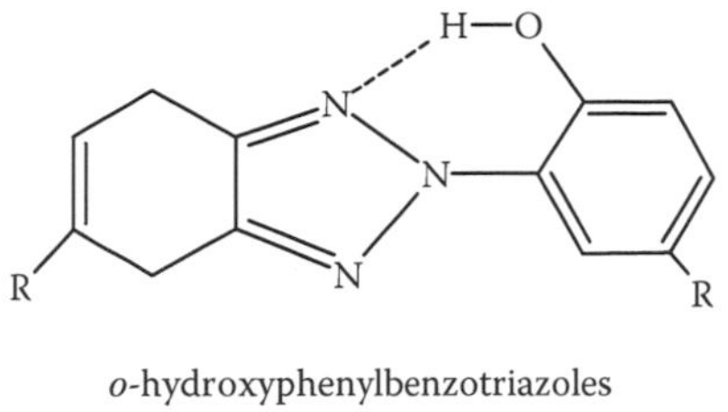

o-hydroxyphenylbenzotriazoles

FIGURE 4.2   Structure of o-hydroxyphenylbenzotriazoles.

o-hydroxyphenyl salicylates

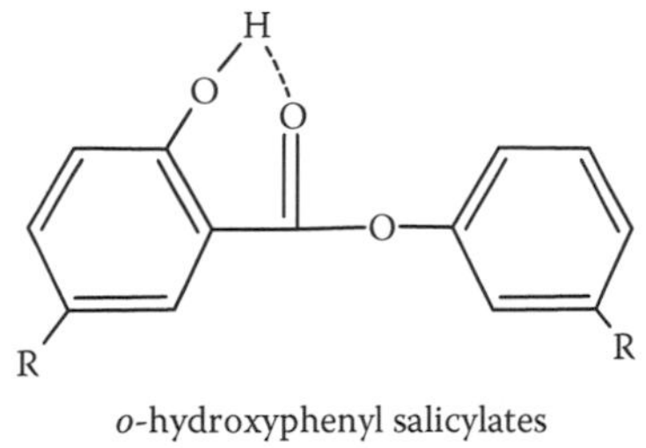

FIGURE 4.3   Structure of o-hydroxyphenyl salicylates.

According to Hawkins, the effectiveness of UV absorbers is severely reduced or eliminated by the following alterations in the molecular structure [3]:

Moving the hydroxyl group from the ortho to the meta or para positions affects hydrogen bonding.

Converting the hydroxyl group to a methyl ether affects hydrogen bonding.

Insertion of one or more CH_2 groups between the carbonyl group and the phenyl ring containing the orthohydroxyl group affects hydrogen bonding.

Replacement of aromatic rings with cyclohexyl rings also affects the hydrogen bond.

Placement of electron-attracting constituents on the aromatic ring containing the hydroxyl group would shift the hydroxyl proton away from the carbonyl group and thus reduce vibration of the hydrogen bond.

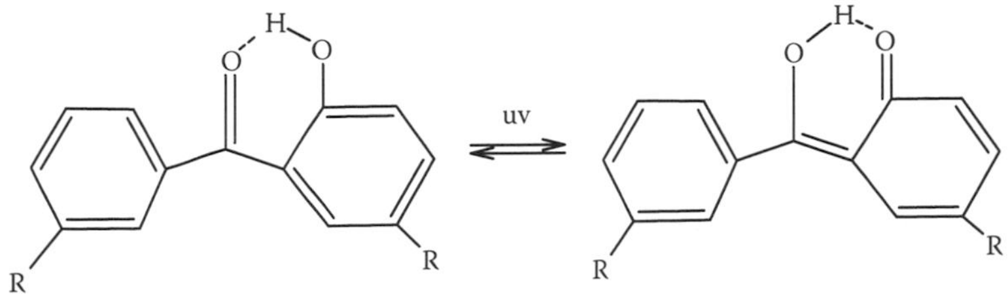

FIGURE 4.4   Keto-enol tautomerism.

Crabtree and Kemp proposed a keto-enol tautomerism as an extension of the mechanism involving vibration at a hydrogen bond [5]. Heat is released in the reverse step in Figure 4.4. This mechanism may supplement the primary mechanism for energy dissipation in UV absorbers.

Radical traps include hindered amines that protect against photodegradation and are called HALS (hindered amine light stabilizers). These are highly efficient molecules that can be used at concentrations as low as 0.1% [3]. The great benefit of utilizing these molecules is that they are as effective as carbon black in inhibiting photodegradation, but can be used without detriment to the appearance of the substrate. These HALS do not absorb UV radiation above 270 nm [6,7]. Thus, they are usually not effective as UV absorbers but must inhibit photodegradation by different reactions.

A mechanism for a reaction between HALS and hydrogen peroxide has been proposed by Scott and co-workers [8]. These hydroperoxides are formed in the polymer and react with the HALS to form stable nitroxyl radicals that are believed to be responsible for the stabilization [3]. Figure 4.5 shows this reaction with the nitroxyl radical, which is a very effective scavenger of alkyl or macroalkyl radical as well as the substituted hydroxylamine formed by this reaction. Substituted hydroxylamines can react with peroxy radicals to regenerate nitroxyl radicals [3].

Quenchers [9–12] deactivate chromophores in excited polymer molecules through an energy-transfer mechanism before these excited states can undergo reactions that would result in the polymer's degradation [3]. This process is shown in Figure 4.6. Here the superscript $^\circ$ represents the ground state and * shows an excited state. Likewise, an excited state complex may be formed and this complex could then

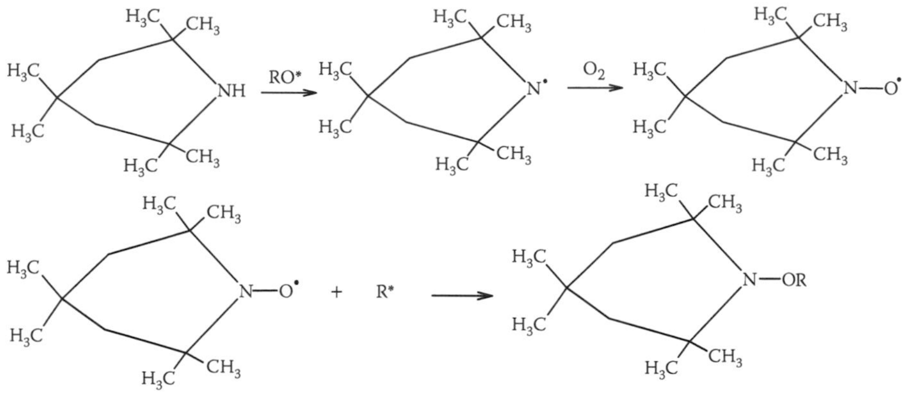

FIGURE 4.5 HALS stabilization reaction.

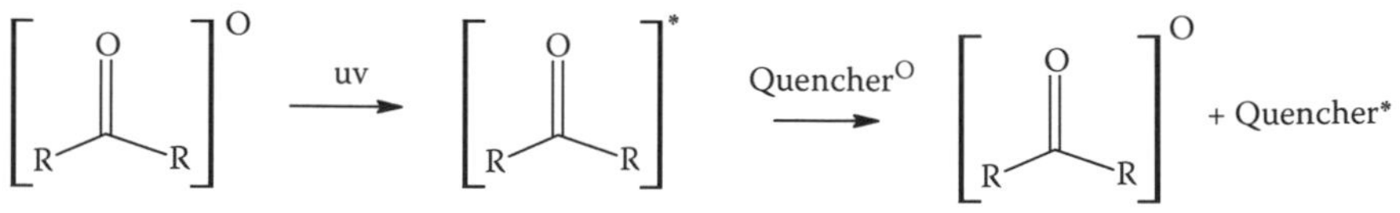

FIGURE 4.6 Ultraviolet quenching mechanism.

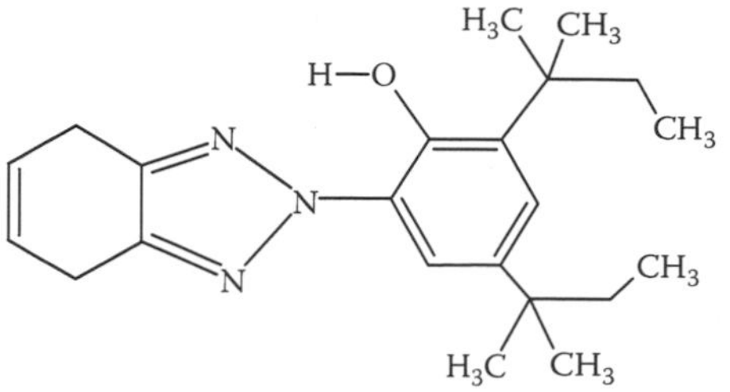

BZT 1
Solvent base: 100% active form
Water base: Aqueous dispersion

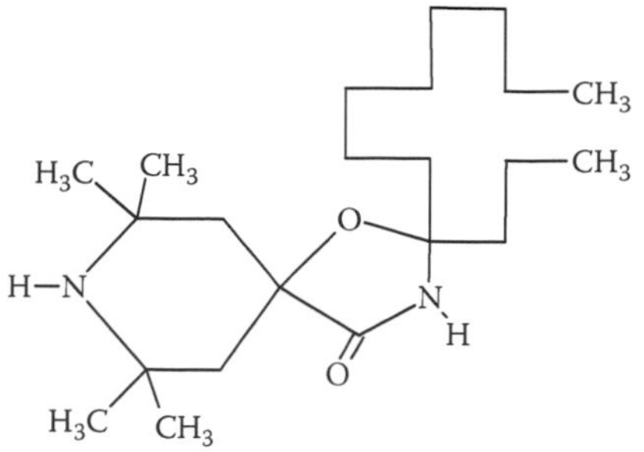

HALS 1
Water base: Aqueous dispersion

FIGURE 4.7 Solvent-based benzotriazole and water-based HALS 1.

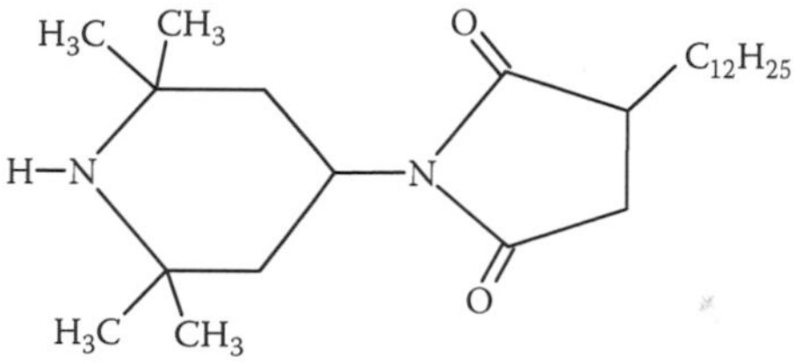

HALS 2
Solvent base: 100% active form

FIGURE 4.8 Solvent-based HALS 2.

dissipate absorbed energy as heat [3]. The main function of quenchers is to extract the absorbed energy from excited polymers before they are degraded. Chelates of transition metals form the most common class of quenchers [3].

Coatings exposed to sunlight for extended periods are also subjected to degradation. Combinations of HALS and ultraviolet light absorbers (UVAs) are used to obtain a synergistic effect. A 1:1 combination of a UVA of the benzotriazole class with HALS has been used [13]. Figures 4.7 and 4.8 show the benzotriazole and HALS combination.

4.4 PREDICTIVE DESIGN TOOLS FOR THE PERFORMANCE IMPERATIVE

Tools for the predictive behavior of a design have developed from classical and numerical methods of the past to the current finite element analysis (FEA) utilized by today's engineers and chemists. FEA is a computer-based analytical tool used to perform stress, vibration, and thermal analysis of mechanical systems and structures. A set of simultaneous equations will represent the behavior of a system or structure under load. Because this is a very important tool, some time will be devoted to the discussion of it, but this is not meant to be a comprehensive study.

Further reading is recommended. Today's chemist should receive some training in computer-aided analysis and design if he or she desires to work in the automobile industry.

4.5 SOME HISTORY OF FINITE ELEMENT ANALYSIS

Historically, we can think of FEA as beginning in 1678 when Robert Hooke determined that force is proportional to a constant multiplied by a strain (Hooke's law):

$$F = kx \qquad (4.1)$$

The evolution of FEA really began in 1940 with Hrennikoff developing truss-based numerical methods to analyze general structures. A timeline would show:

1943: Robert Courant proposes breaking a continuous system into triangular elements.

1950: Boeing uses analog computers, triangular elements, and matrices to analyze surfaces.

1960: Dr. Ray Clough coins the term *finite element*—an entity that can model three-dimensional strain.

1962: MacNeal–Schwendler Corp. (MSC) develops a general purpose FEA capability for NASA, dubbed NASTRAN.

1970: MSC creates enhanced MSC/NASTRAN.

1975: ANSYS, MARC, and SAP are introduced.

1980: CAD capabilities improve, allowing graphical "pre" and "post" output processors for FEA tools.

1998: FEM/A (FEA software enable) software enables the nonspecialist to perform various complex analyses up to 1 million degrees of freedom.

Computer-aided design provides an ideal opportunity for the design team to make predictions about the behavior of a design. The chemist or engineer can make a computer model of a part utilizing the proper inputs and have prior accurate knowledge of material choice that will generate the desired performance. This approach will also affect the cost imperative because costly physical testing is eliminated. Table 4.2 shows a comparison of testing versus finite element analysis. Physical samples are not required; energy for running machines and manpower for testing are also saved.

The outputs from finite element analysis will tell the chemist many of the important predictive characteristics of the part. The effects of load (stress, strain, and deflection) are measured. Material selection based on strength and part dimensions and tolerances are determined and easily adjusted. The effects of heat (heat transfer and thermal stresses) are determined as well as material selection based on conduction and insulation. Special applications such as bending and vibration characteristics, crashworthiness, fatigue, and noise can be determined. The design requirements, such as expected loads, load cycle cost, mass targets, and budget targets, can be predicted and met with proper application of the FEA tool.

TABLE 4.2
Finite Element Analysis versus Testing

Finite Element Analysis	Testing
Building FE model and running analysis are cheaper and faster	Building and testing physical parts is costly and takes a long time
No parts are physically involved	Tests are usually destructive
Multiple analysis must be done to account for part and build variations	Multiple physical tests must be done to account for part and build variations
Evaluation of different design alternatives is very easy	Evaluation of different design alternatives is a long and costly process
Design optimization is easier	Design optimization cannot be done easily

4.6 FEA PERFORMANCE PREDICTIONS AND SOME KEY DEFINITIONS

Accuracy: The ability of the FEA to predict actual behavior, mainly affected by the quality of model inputs, mesh refinement, and others.

Convergence: Technique for obtaining greater accuracy from an FEA by iteratively refining the mesh in areas of greatest concern until outputs stabilize.

Isotropic: In FEA, this concept is assumed. It is a material that has the same strength in all directions (as opposed to anisotropic materials).

Modulus of elasticity (E): Also known as Young's modulus, a numerical constant unique to each material that describes the elastic properties of the material when loaded in the compression or in tension.

Poisson ratio: The ratio of transverse contraction strain to longitudinal extension strain in a stretched bar. The maximum possible value is 0.5 (around 0.5 for rubber, 0.33 for aluminum, 0.28 for common steels, and 0.1–0.4 for polymer foams).

Singularity: Geometrically, the term used to describe a discontinuity such as a fillet or part wall that causes an FEA result to diverge as the mesh is refined.

Stiffness: A measure of the force required to deform a material in the elastic range. To increase stiffness, a chemist or engineer will use a material with a higher Young's modulus.

Strain: Ratio of the change in length of a part to its original length ($\Delta l/L$).

Strength: A measure of a material's ability to resist loads without deforming permanently. To increase strength, a material with a higher yield stress must be used.

Stress: Force per unit area in newtons per square meter (pascal) or pounds per square inch. If the material is stretched, then it is in tension. Compressive stress is often examined as well.

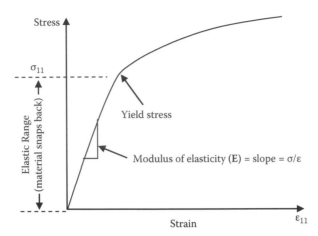

Stress–strain curve: Graphical representation of the stress versus strain in a part as the part goes from an unloaded to a fully loaded state.

Toughness: A measure of a material's ability to absorb energy in the plastic range. It is measured by the material's strain energy density or the area under the stress–strain curve up to the maximum stress.

Yield stress: Stress level at which permanent material deformation begins.

FEA is fundamentally based on static equilibrium principles. The sum of all forces in any direction and the sum of all moments about any axis are zero. In a static analysis, the dynamic aspect of the load is ignored. As in static analysis, a free body diagram (FBD) is used. An FBD gives a picture of the part with the loads and moments represented graphically. For any point, there are six degrees of freedom: three translations along each axis and three rotations about each axis. In static equilibrium, the sum of all forces in any direction and the sum of all moments about any axis are zero. In static analysis the dynamic aspect of the load is ignored.

FBDs are important in analysis because they allow visualization of the things that are felt by the part but not shown on the screen. FEA has its roots in truss analysis, where each member of the truss carries only axial load transmitted from someplace else on the truss. The free body diagram is a sketch of each element in isolation, showing force arrows where they occur at the pins and supports. The equilibrium equation is

$$\sum F_x = \sum F_y = 0$$

A simple loaded truss is shown in Figure 4.9.

The free body diagram will break down these truss models into its individual members and analyze which type of force or moment is acting upon each member at each node, or point. Figure 4.10 shows the FBD of the simple loaded truss; it shows if the member is in compression or tension (stretching) and shows the force designated (P) on each member.

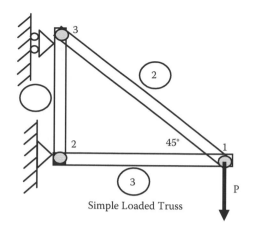

FIGURE 4.9   FBD of simple loaded truss.

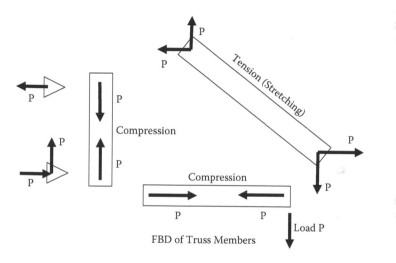

FIGURE 4.10   FBD showing tension and compression of truss members.

The free body diagram is considered for the entire structure and each node is examined. The FEA model breaks a structure into smaller elements, each representing the material's stiffness. Load transfer is calculated through smaller elements and by considering equilibrium, compatibility of deformation, and Hooke's law. In Figure 4.11, a load is placed on a fixed beam as shown. In the three-element representation, each element is affected by the initial downward and axial loads.

As a measure of stiffness, the Young's modulus is important in the predictive behavior of the material being used. For linear analysis, E = stress/strain. For automotive applications, some common materials are steel ($E \sim 200$ GPa), aluminum ($E \sim 70$ GPa), and nylon ($E \sim 8.5$ GPa). As stated earlier, Hook's law is force = spring constant * spring displacement ($F = KU$). The generalized Hook's

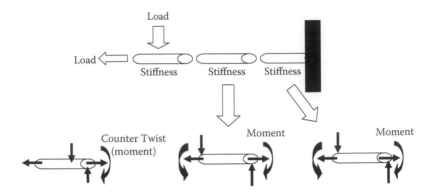

FIGURE 4.11   FBD moments.

law states that stress = modulus of elasticity * strain ($\sigma = E\varepsilon$). For a beam in tension,

$$\sigma = F/A \text{ and } \varepsilon = \Delta L/L$$

$$F = (EA/L)\Delta L; \text{ since } F = KU, \text{ then}$$

$$K = EA/L \tag{4.2}$$

The chemist or design engineer can make material choices (material type, filler, etc.), design adjustments (thickness of material, whether the part needs ribs or gussets, etc.) based on this predictive tool.

When an FEA model is run, several elements must be present. These include the CAD data, material properties, loads acting on the part, and the boundary conditions used. Table 4.3 shows the typical input to conduct an analysis through one of the software programs. The minimum input for structural analysis is the modulus of elasticity, Poisson ratio, and density. For thermal predictions, the minimum inputs are coefficient of thermal expansion, specific heat, and thermal conductivity. For modal analysis, the minimum inputs are modulus of elasticity, Poisson ratio, and density.

TABLE 4.3
Thirty-three Percent GF Nylon Properties

Young's modulus	8487 MPa
Poisson's ratio	0.35
Density	1.34 kg/mm³
Thermal expansion	4.3e-5 1/°C
Tensile yield strength	141 MPa
Tensile ultimate strength	141 MPa
Compressive yield	138 MPa

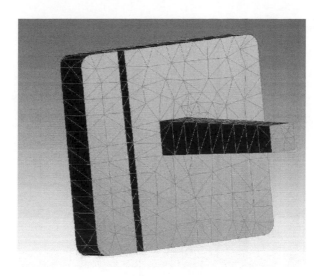

FIGURE 4.12 Part with cantilevered beam showing FEA nodes.

Figure 4.12 shows a nylon 6,6 beam with the properties from Table 4.3. The figure shows math data (CAD data) that was imported into the FEA program and meshed. Meshing is the process of dividing the structural volume or area of a model into a finite number of smaller numbers of elements. Some software has automeshers and others require a user to mesh utilizing a separate program. A good mesh will conform to the geometry of the part, have smaller elements where geometry is more important and larger elements where geometry is less important, and be free of defects. Figure 4.13 shows the various types of elements used in meshing.

Solid elements should be used for parts with thicknesses greater than 3 mm. Shell elements are used for sheet metal parts (hoods, deck lids, doors, etc.). A course mesh means that the computer-generated solution will be faster but less accurate and use less computer memory. A finer mesh means exactly the opposite—a slower, more accurate solution. In our fictional part, Figure 4.14 shows the application of a 500-N

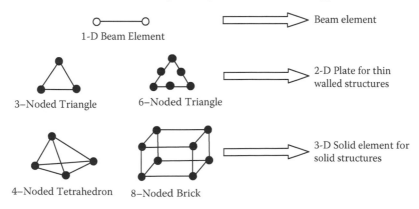

FIGURE 4.13 Diagram of node types and uses.

FIGURE 4.14   Part with cantilevered beam showing force application.

force on the end of the tapered beam element. The part also has a boundary condition fixing it to a stationary position. A thermal condition of 100°C was also added to the analysis. Thermal loads allow temperature-dependent convection from a chart of temperature versus film coefficient input from a table or by the user. Each of the surfaces needs to be designated as to where the temperature is applied. The program calculates heat flux (q/a) into or out of the surface:

$$q/a = h(t_s - t_f) \tag{4.3}$$

Here, h is the heat transfer coefficient, t_s is the temperature on the surface, and t_f is the bulk fluid temperature. The thermal load was placed to simulate an operating condition of an automobile part.

A black and white representation is shown here, however, real outputs of the various programs are given in color with an accompanying chart. The outputs of the various programs are given in color with an accompanying chart. These plots are called contour plots. Typically, red will designate the maximum condition being examined and blue or green will indicate a minimum. Figure 4.15 shows the stress plot of the output with a maximum stress condition showing 125 MPa (N/mm²) von Mises. Stress is calculated for each element and then averaged at the nodes to create a contour plot. A tag will show the chemist or engineer the maximum as well as the minimum stresses. The von Mises stress is most appropriate for ductile materials. von Mises combines the three stresses (x, y, and z) and compares them to the yield stress of the material. It is important for the chemist to keep the stress of a part below the yield of a material by a certain safety factor.

Another important performance imperative is NVH (noise, vibration, and harshness) concerns. FEA can help in this aspect as well by making predictions about natural frequencies. In our example, the frequency predicted for the first mode is

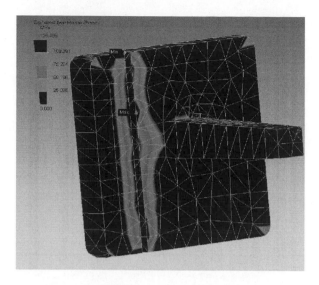

FIGURED 4.15   Part with cantilevered beam showing stress plot.

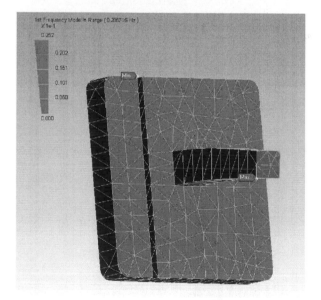

FIGURE 4.16   Part with cantilevered beam showing frequency plot.

0.25 Hz (shown in Figure 4.16). This is the frequency at which our beam will vibrate when the 500-N force is applied at the indicated position.

In automotive applications, it is extremely important for engineers from different groups to work in conjunction with each other to avoid conflicting natural frequencies (a condenser radiator fan module vibrating at 75 Hz would not want to conflict with a steering column vibrating at the same frequency). This condition would cause customer dissatisfaction by amplifying the vibrations and increasing NVH. A beam can be excited in various patterns at different frequencies. This property is an indication of the part's stiffness and is proportional to the square of the stiffness divided by the mass or

$$f = \sqrt{\frac{stiffness}{mass}}$$

Another very important predictive output is the amount of deformation that the material is subject to. These are typically called total displacement plots, as shown in Figure 4.17. In our example, the beam was displaced 2.003 mm.

4.7   PREDICTIVE DESIGN FOR THE COST IMPERATIVE

Optimizing design for the cost imperative is critical for today's applications. Performance to cost design is affected by a number of factors. These factors include material cost, processing cost, tooling cost, secondary cost, and inventory cost. The simple spreadsheet shown in Figure 4.18 can be used as a guide to help understand and track raw material cost. Typically, the chemist would approximate the length, width, and height to get an approximate volume for the part. The material quantity can then be calculated from the part's volume and density. The material quantity is corrected for material scrap loss. The price per pound or kilogram minus overhead will then give the total material cost.

Manufacturing cost can also be comprehended by the use of a spreadsheet and some knowledge of the process. Figure 4.19 shows manufacturing costs that should be considered. The equipment purchase price is considered along with the amoritization rate, number of shifts, interest rate on the equipment, energy cost for the equipment, and floor space cost. Equipment time rate should be considered. The equipment time

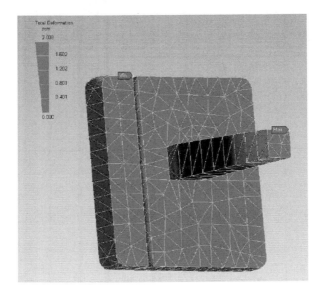

FIGURE 4.17   Part with cantilevered beam showing total displacements.

Part Number

Program		Total Cost:	0.805 $/Unit		(1X = .805 $)	

Part Number
Tier 1 Supplier: **Supplier X**
Tier 1 Location: **Supplier location**
             Tier 1 or 2:

Material	Manuf Equip	Manuf Lab	SG&A + Profit	Others
0.242	0.376	0.026	0.161	

Data:
Quantity
Part Name
Supplier Part Number
Drawing Position
Supplier
Supplier Location

Net Quantity of material	318 g
Raw Material	Carbon Steel 1008
Material Scrap & Losses	5%
Material Overheads	5%

Machine	Deep Drawn Press (for Motor Housing)
Cycle Time	10 s
Number of Pieces out/Cycle	0.0
Uptime	85.00%
Manufacturing Scrap	1.00%
Number of Shifts	2
Workers per machine	2
SG&A, R&D+Profit Material	25%

Material Cost

Length (mm)	Width (mm)	Thickness (mm)	Material Densisty (Kg/dm^3)	Neq Qty Material (Kg)	Material Scrap & Losses	Gross Qty Material (Kg)	Price ($/Kg)	Material Overheads (%)	Total Mat Cost ($/Piece)
102	102	1.65		0.318	5.00%	0.3339	0.69	5.00%	$0.242

FIGURE 4.18 The beginning of a spread sheet indicating cost breakdown.

rate is the amortization added to the interest rate, maintenance, tool maintenance, energy, and insurance. Labor rates (direct and indirect) should also be factored in. Total time rate (equipment and labor) and cycle considerations, including uptime and scrap rate, are also shown in Figure 4.19. A summary sheet (Figure 4.19) will include the raw material cost; manufacturing and assembly cost; selling, general and administrative (SG&A) expenses and profit, overhead; and research and development.

This is a useful and powerful tool for the chemist and design engineer. Usually, with the help of purchasing or the supplier assistance arm or purchasing (supplier quality or development), this type of information can be acquired when a design is being prepared. At that point, the chemist or design engineer can consider how to best affect the cost of the part. Some of the ways to do this in the design phase are shown in Table 4.4.

Cycle time improvements can be made and thus affect cost with the design of the part. Tools such as a mold flow analysis will predict how the part will behave in the mold. The designing engineer can make changes to the part that can improve cycle time, and he or she can test them virtually using FEA and mold flow analysis. The tooling cost can be affected by utilizing a different type of filler material, which may facilitate the use of an aluminum tool or a P-20 steel tool versus a hardened or H13 tool. Slides and cams can be removed by the use of snap fit connections or other design assembly schemes. Secondary cost, such as painting, can be minimized by choosing a paint or coating type that will meet specifications but be cost effective (i.e., epoxy vs. polyester, or E-coat vs. powder coat). Minimizing the number of parts by combining functions within a single part will also greatly affect the cost of the part.

MANUFACTURING COST

Equipment Purchasing Current Price				$	1860000
Amortization Time				yrs	7
	1 Shifts	2 Shifts	3 Shifts		2 shifts
	1740h	3480h	5220h		3480
Interest Rate					7%
Interest Calculated on					Life Average
Maintenance Percentage of Investment					1.40%
Installed Power				kw %	100
Energy Cost		Electricity		$/kw	0.14
Average Energy Consumption Ratio					50%
Equipment Dimensions				Length (m)	7.62
				Width (m)	7.62
			Used sides of machine		1 Sides
			Evolution Ration (from 1 to 2)		1
Area Cost				$/m^2	31.76
Equipment Area			(4 time LxW)	m^2	232.2576
Insurance Ratio					1%

EQUIPMENT TIME RATE

1-AMORTIZATION	$/hour	76.35
2-INTEREST RATE	$/hour	18.70
3-MAINTENANCE	$/hour	3.88
	Tooling Cost ($)	
4-TOOLS MAINTENANCE	$286,000.00 $/hour	0.60
5-ENERGY	$/hour	7.00
6-AREA	$/hour	2.12
7-INSURANCES	$/hour	5.35
TOTAL EQUIPMENT TIME RATE	$/hour	114.00

FIGURE 4.19A Spreadsheet with manufacturing cost per unit.

4.8 STRUCTURAL DESIGN CONCERNS

A key in designing for part performance is stiffness of the part. There are a number of ways to optimize the stiffness of a part. The design engineer can use a higher modulus material or he or she can make the part thicker; the engineer can change the filler type or the design can be changed to incorporate a stiffener such as a rib or gusset.

If material is located away from a neutral axis, the part can be stiffened. This technique increases the moment of inertia of the cross section [14]. Basically, this relationship is

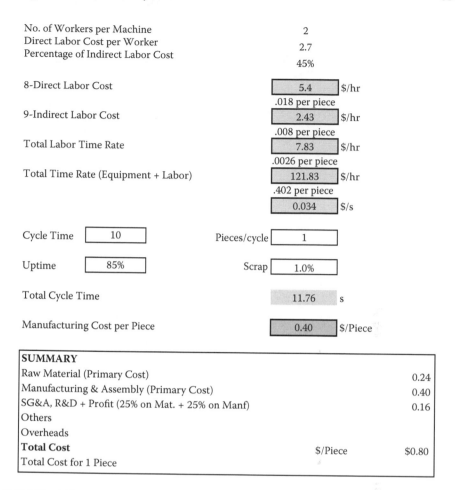

No. of Workers per Machine	2	
Direct Labor Cost per Worker	2.7	
Percentage of Indirect Labor Cost	45%	

8-Direct Labor Cost	5.4	$/hr
	.018 per piece	
9-Indirect Labor Cost	2.43	$/hr
	.008 per piece	
Total Labor Time Rate	7.83	$/hr
	.0026 per piece	
Total Time Rate (Equipment + Labor)	121.83	$/hr
	.402 per piece	
	0.034	$/s

Cycle Time	10	Pieces/cycle	1
Uptime	85%	Scrap	1.0%

Total Cycle Time	11.76	s
Manufacturing Cost per Piece	0.40	$/Piece

SUMMARY		
Raw Material (Primary Cost)		0.24
Manufacturing & Assembly (Primary Cost)		0.40
SG&A, R&D + Profit (25% on Mat. + 25% on Manf)		0.16
Others		
Overheads		
Total Cost	$/Piece	$0.80
Total Cost for 1 Piece		

FIGURE 4.19B   Cost due to labor and summary.

$$I = \int y^2 da$$

(4.4)

The advantage of this type of stiffening is that it contributes more (a squared function vs. linear for a modulus change) from a design perspective as well as a cost perspective. Design changes save material cost versus making changes to material, filler, and thickness.

Ribs are the most common design stiffener. They will change the stress pattern and reduce the amount of deflection of a part. Ribs are oriented along the axis of bending to provide maximum stiffness [14] in a simply supported part. If torsional loads are applied to a part, then a diagonal rib pattern would be desired. The design engineer must balance strengthening versus aesthetics when adding ribs.

TABLE 4.4
Opportunities for Cost Reductions

Processing cost	Cycle time
	Number of cavities
	Operator required
	Size and type of press
	Ability to use regrind
Tooling cost	Can aluminum be used?
	Slides and cams
	Dimensional concerns
Secondary cost	Painting or shielding
	Assembly required
	Fasteners required
Inventory cost	How many parts?
	Easy to handle (stackable)

Gussets are usually used to reduce localized regions with large deflections. A gusset is basically a triangular brace that will transfer a load from one section of the part to another. Figure 4.20 shows a gusseted part.

A design engineer must consider tooling implications when designing gussets and ribs. As a tool comes apart, the ribbing must be in the direction of draw. When a flat surface is encountered, a corrugated surface can be used. With these surfaces, additional material is not needed. Another benefit is that no additional cooling time is required for the part [14]. Stiffness is increased only in the direction of orientation, however. As with ribs, stiffness is obtained by increasing the average distance of material from the neutral axis of the part, which increases the moment of inertia [14]. A corrugation is a very effective stiffener; however, it is difficult to obtain even surfaces and proper tolerance control.

4.9 STRENGTH AND IMPACT CONCERNS FOR PERFORMANCE

Designing for strength in a part can be thought of as designing the part to be able to withstand the application of the maximum load applied to a part before the part fails to function as intended [14]. General Electric defines common modes of failure as fracture, permanent deformation, and inadequate stiffness [14]. With fracture, the stresses in the part are more than the ultimate strength of the material and a break will occur. With permanent deformation, the load will result in stress in the part that exceeds the yield point of the material. Deflection or strain that is not fully recoverable is produced; this will result in a distorted part. If a part is not stiff enough, the load imparted on the part will deflect it to affect the function of the part. Some concepts for part strength should be considered [14]:

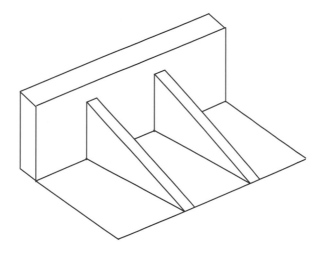

FIGURE 4.20 A gusseted part.

Part yield strength. The part is loaded beyond the normal functioning range of loads and the onset of permanent deformation beyond a certain point determines the strength of the part. This deformation can cause the part to stop functioning.

Part stiffness. When a part is loaded, the amount of deflection that occurs is considered to be the strength of the part. The ratio of load to deflection is actually the stiffness of the part rather than the strength of the part.

Strain-controlled applications. In these applications, a definite strain is applied. The failure mode for the part may be fracture from exceeding the maximum elongation for the material, permanent deformation from exceeding the yield point, or fracture over time from strain close to the maximum elongation and subsequent creep rupture.

Part ultimate deflection. The maximum deflection of an application is dependent upon the percent strain to failure of the material, part geometry, loading, and constraint.

Part yield deflection. The part is loaded beyond the normal functioning range of deflections and the onset of permanent deformation determines the strength of the part. This permanent deformation causes the part to malfunction.

Part toughness. Toughness is measured by calculating the area under the load-deflection curve, from initial load to part failure. The toughness of the part is applicable for both load- and deflection-controlled applications.

When the part is being designed, stress concentrators should be avoided. These are items such as sharp notches, internal corners, and sharply angled wall sections. Surface interruptions such as holes and inserts can significantly reduce the strength of plastic parts and induce failure. When parts are designed for performance, certain questions should be asked:

How can an improper installation affect part life?
What is the highest temperature the part will face?
What are the properties of the material at that temperature?
What chemicals will come in contact with the part?
What are the maximum loads the part will see?

Answering these questions correctly will lead to a good design for a vehicle that performs to design standards at a satisfactory cost.

REFERENCES

1. http://media.gmcanada.com/division/chevrolet/products/archive_prod_info/ /99chevy/overview/story/bod_1931.htm
2. Guggenheim and Martin. *Handbook of Clay Science.* 1995. 255–256.
3. Hawkings, W. L. 1984. *Polymer degradation and stabilization,* 18–19, 75, 78–84. Berlin: Springer–Verlag.
4. Hirt, R. C., N. Z. Searle, and R. G. Schmitt. 1961. Ultraviolet degredation of plastics and use of UV absorbers. *Society of Plastics Engineers Transactions* 1:21.
5. Crabtree, J., and A. R. Kemp. 1946. Accelerated ozone weathering test for rubber. *Industrial and Engineering Chemistry* (analytical ed.) 18:769.
6. Huisgen, R. 1963. 1,3 Dipolar cycloadditions, past and future. *Angewandte Chemie* (international ed.) 2:565.
7. Criegee, R., and G. Schjroder. 1960. Direct observation of polymerization in the oleic acid-ozone heterogeneous reaction system by photo electron resonance capture ionization areal mass spectrometry. *Chemische Berichte* 93:689.
8. Murray, R. W., R. D. Youssefyeh, and P. R. Story. 1967. Mechanism of photo sensitized oxidation. *Journal of the American Chemical Society* 89:2429.
9. Privett, O. S., and E. C. Nickell. 1963. Determination of structure of unsaturated fatty acids via reductive ozomolysis. *Journal of the American Chemical Society* 40:22.
10. Loan, L. D., R. W. Murray, and P. R. Story. 1965. The mechanism of ozomolysis formation of cross ozonides. *Journal of the American Chemical Society* 87:737.
11. Murray, R. W., R. D. Youssefyeh, and P. R. Story. 1966. The unequivocal ozonide stereoisomer assignment. *Journal of the American Chemical Society* 88:3143.
12. Biggs, B. S. 1958. *Rubber Chemistry and Technology* 31:1015.
13. Bechtold, K., and H. Clariant. 2006. *Material Testing Product and Technology News* 36:4.
14. GE Engineering Thermoplastics. *Design guide,* 3–34. Pittsfield, MA: General Electric Company.

5 Manufacturing and Process Technology

5.1 INTRODUCTION

This book was written from the perspective of the OEM (original equipment manufacturer) chemist. Thus, manufacturing or processing for various plastics, elastomers, or metals is not discussed in detail, although some time should be devoted to it. Often, the OEM chemist or material engineer is required to interface with raw material suppliers and chemists from tier one or tier two suppliers regarding processing and how best to manufacture the product of interest. We will briefly touch on some of the most important and useful technologies for the chemist today. The discussion centers on some of the key processes and technologies within this book.

5.2 RUBBER PROCESSING

Rubber processing can be considered to take place in four steps: mixing, forming, storing, and molding. During the mixing process, heat applied to the rubber will cause the rubber to go through various states of flow. Below the T_g (the glassy region), the polymer chains are frozen in place. The material is highly elastic and molecular motion is limited to bending and stretching. At the glass transition region (T_g), rotational motion increases and the polymer is very viscous. In the rubbery region, the polymer is highly elastic and rubber like, which will allow for elongation. In this region, chains develop entanglements that act like cross-links. Finally, in the flow region, at the highest temperatures, resistance to translational motion decreases and the chains slip easily past one another. Viscous flow is now possible, dimensional stability is lost, and the material can be compounded using mixers, rubber calendar machines, and curing presses. Finished products will no longer flow because they are now cross-linked.

Rubbers must first be mixed. They consist of various components that provide a specific function to the compound. A typical rubber compound has the components listed in Table 5.1. To mix the components, a Banbury mixer is often used. These types of mixers have been in use since 1916 [1]. A Banbury will have two spiral blades held within a housing and is considered an internal mixer. The two blades of the mixer will have channels for heating and cooling if necessary. Between the blades is a ridge or space for the material to be mixed. The invention of the Banbury resulted in the removal of roller milling in rubber production [2]. Figure 5.1 shows a rough schematic of the internal Banbury type mixer. For a typical tire batch of 225 kg, a Banbury mixer has a gross capacity of around 270 L and operates at up to 2,000 hp [3].

TABLE 5.1

Rubber Components and Their Functions

Component	Function
Elastomer	Base
Processing aids	Processing help
Vulcanizer	Cross-linker
Accelerators	Increases cure rate
Activators	Modifies cure system
Stabilizers	Diminishes degradation
Fillers	Reinforcement
Softeners	Mixing aid

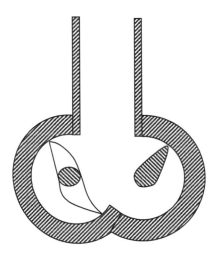

FIGURE 5.1   Cross section of internal mixer.

If a smooth finish is desired, a calender is used to process at the end of the processing cycle. A calender is a set of pressurized rollers usually made of chilled cast iron but sometimes steel. The calender is heated and operates in a continuous manner; however, it is usually employed with vinyls and ABS polymer sheets.

When designing a compound, the chemist considers several factors in line with design imperatives:

- cost (cost of processing, raw materials, etc.);
- fabrication (best process for desired characteristics);
- engineering properties desired (blend for modulus, maximum use temperature, creep, stress, etc.)
- performance properties desired (blend for tensile, elongation, fatigue, etc.); and
- environmental considerations (blend for oxidation resistance, oils and solvents, etc.).

Table 5.2 shows the typical compositions of some of the compounds used in seal and gasket applications (Chapter 8).

The bulk of the cost of raw materials is the base polymer itself at approximately 13%, depending upon the blend. Extrusion (as well as compression molding) and curing will add the bulk of cost to a polymer at around 40% (see Figure 5.2). When a material is extruded, it is forced through a die of fixed cross-sectional area, often heated while a continued cure or vulcanization takes place.

The cost from extrusion is due to the capital cost incurred for the machine and dies, and the energy cost to run the ram. Machines can cost upwards of $100,000, and dies can exceed $5,000. When a rubber is vulcanized, sulfur bridges connect individual polymer units, making the overall compound harder and more resistant to chemical attack. This is an irreversible process that creates a thermoset material. For automotive applications, tires are vulcanized and compression molded more than any other component. Figure 5.3 shows this process with the addition of heat.

This type of thermoset curing system is advantageous for tires and components that require irreversible heat and chemical-resistant properties. Peroxide can also be used as a curing system. In the sulfur-cured system, the sulfur atom attacks the allylic hydrogen atom at the cure site and forms a two- to eight-member cross-link. A shorter sulfur cross-link will give increased thermal properties; a longer cross-linked system will give better dynamic properties. Accelerators and activators such as stearic acid are all mixed in a cure package and added to the mixing chamber. During compression molding for SBR tires, for instance, a temperature of 170°C is applied for about 10 minutes.

TABLE 5.2
Parts by Weight of a Typical Rubber

Component	Diene	EPDM[a]	NBR[b]
Base	100	100	100
Filler	0–80	40–200	40–80
Plasticizer	0–40	0–200	0–30
Process aid	0–5	0–5	0–5
Stabilizer	1–3	—	1–5
Wax	0–4	—	0–3
Lubricant (fatty acid)	1–2	1–2	1–3
Activator (ZnO)	3–10	3–10	3–10
Vulcanizer	1–3	1–2	1.5–4
Accelerator	0.5–2	1.5–3	1.5–2

Note: Parts by weight of typical rubber.

[a] Ethylene–propylene diene monomer.

[b] Nitrile rubber.

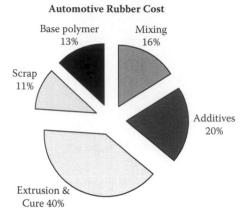

FIGURE 5.2 Cost breakdown of automotive rubber.

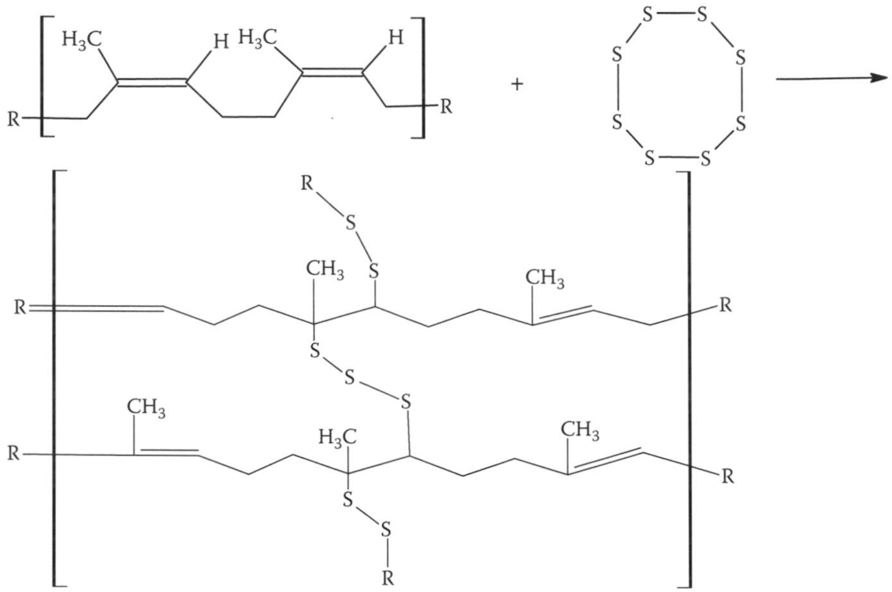

FIGURE 5.3 Vulcanization of rubber.

5.3 PLASTIC PROCESSING

The main processing technique in the automobile industry is injection molding in which pressure is applied to make the polymer flow into a preferred form [3]. Plastics must be cooled below the T_m (melt transition temperature) before removal from the mold [3]. The molding process follows a basic injection cycle consisting of mold close, injection carriage forward, injection of material, metering, retraction of carriage, mold open, and part ejection. This cycle has been in place basically since 1946, when James Hendry introduced the screw injection molding machine [4]. Newer electric injection molding machines have been replacing the hydraulic machines and have increased the cycle's repeatability and reproducibility at a lower operational cost. Much of the success of part production lies in designing a proper tool. For instance, cooling in the tool is critical. If a mold is not capable of removing heat in critical areas of the mold, distortions can occur. Likewise, parting lines, ejector pins, gates, and runners must also be designed correctly. Table 5.3 is a list of partial molding issues and their common causes.

Raw polymer is fed into the injection machine in pellet or granule form. These pellets are fed into a hopper connected to the end of a cylindrical barrel. Injection molding will accomplish its cycle in separate zones within the apparatus [3]. A

TABLE 5.3
Common Part Defects and Their Causes

Defect	Description	Cause
Blister	Part surface blemish	Tool is too hot
Gas burn	Burned area on part	Vent issue or inject speed too fast
Flash	Thin layer of material at or around parting lines	Clamp force too low; damage to tool; excess material
Delaminating	Peeling of thin material on walls of part	Part contamination
Streaking	Local streaking of color	Improper mixing
Jetting	Part deformed by turbulent flow	Too fast injection speed; gate improperly located
Flow marks	Waves in material	Injection speed too slow
Contamination	Foreign objects in part	Contamination in barrel or tool surface
Warping	Unevenness or flatness within the part	Cooling cycle not long enough; material too hot; lack of cooling due to poor cooling design
Voids	Holes within the part	Improperly pressurized cycle; lack of holding pressure
Short shot	Mold not filled completely	Inject pressure too low; lack of material
Splay or streaking	Patterns around gating	Material damped prior to injection
Sink	Sink marks in part	Insufficient pressure or hold time, or cooling time too low
Prominent knit line	Flow front very prominent	Tool or material temperature too low
Degradation	Hydrolysis of material	Raw material inadequately dried; water in material or barrel temperature too low

piston or plunger will push the granules into the hopper when the barrel is withdrawn into a heated barrel. A screw within the barrel is rotated thereby feeding the pellets through the screw's grooves. The barrel has heated zones that assist in the process; however, a great deal of heat and pressure is generated as the screw flights decrease toward the end of the screw toward the mold. This mechanical action will heat the plastic as the screw rotates.

The sprue is the channel from which the resin will flow through the gate and into the mold. Runners are perpendicular to the direction of draw, and the sprue is parallel to the direction of draw. This portion of the system is often recycled because typical OEMs allow up to 15% of regrind material (the use of polymer salvaged from the runner or sprue and not used in the original part). When injection molding takes place, automotive thermoplastics must be cooled under the T_m or T_g before ejection.

Injection mold machines are rated by their ability to mold polystyrene in a single shot [3]. Mold pressures can range from 55 to 275 MPa and cycles as low as 15 s. It is possible to injection mold a thermoset resin, although it is very difficult. Compression molding or transfer molding is used in these cases:

- Compression molding is an older method of polymer processing that is still currently used for parts such as polyester thermoset engine covers or poly-amide transmission seals. The resin will be placed in half of the mold (gen-erally the female half shown in Figure 5.4). The male half of the press will compress the resin to a pressure of about 15 MPa. The resin will be heated simultaneously in both halves of the mold, which will cause it to begin to cross-link.
- Transfer molding combines injection molding and compression molding.
- Extrusion is a method of producing a length of material with a uniform cross section [5]. Extruders can be heated and cooled as required. A screw provides pressure to the material after it is fed into the barrel through the feed hopper. The pellets are melted and pass through a breaker plate into the die. The material is then forced out of the die; its cross section is deter-mined by the shape of the die [5].

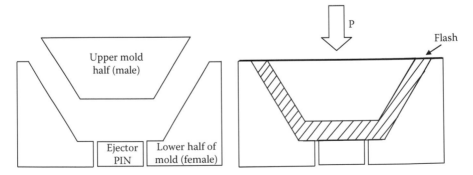

FIGURE 5.4   Compression-molding diagram.

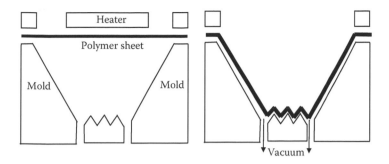

FIGURE 5.5   Vacuum-forming diagram.

- In blow molding, the end of a piece of polymer that was initially formed by an extrusion process is scaled by the closing of the mold. Compressed air is passed into the tube and expanded to fit the mold.
- Vacuum forming (shown in Figure 5.5) is the reverse of blow molding. A sheet of heated material is placed over a mold and the air is sucked from the mold [5]. The plastic is drawn down to conform to the shape of the part, such as a gas tank. Gas tanks comprise several different polymers and are blow molded as sheets. Parts are allowed to cool before being released from the mold.
- Calendering is a process in which a preheated polymer mix is turned into a continuous sheet by being passed through two heated rolls, which squeeze the material to a programmed thickness [5].

A recent development in injection molding is in-mold assembly (IMA). This process is very useful for instrument panel air registers in the HVAC and interior areas. Over 70% of IP air registers have what is known as a dual vane construction—the ability to direct air in two directions. For these applications, a slider knob to control the airflow is necessary as well as dual colors for high-end vehicles. If an IP register can be molded in color as well as assembled in the mold, a cost savings will be realized by eliminating the assembly process as well as a painting operation or coating process. Basically, in the process all components are molded in parallel within one cycle, and the components are moveable. A tool configuration for IMA is shown in Figure 5.6.

In IMA, different types of materials can be injected into a single tool that rotates. For instance, a polyamide material can be injected into injection port 1; injection area 2 can be PBT or some other compound and injection port 3 can be a third material (a different polyamide, perhaps). Cycle times for this process are typically in line with regular injection molding; however, a savings in time is realized because a complete part is ejected from the machine.

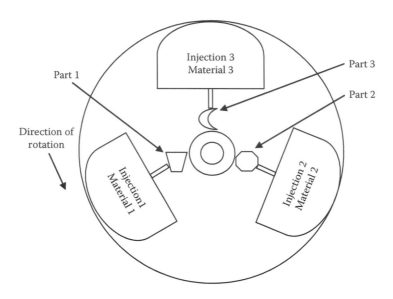

FIGURE 5.6   In-mold assembly diagram.

5.4   ALUMINUM PROCESSING

Aluminum processing is very important in automotive applications due to its light weight and versatility. One of the areas where a chemist's knowledge is used is in heat exchanger technology (i.e., radiators, condensers, heater cores, charge air coolers, and evaporator cores). Aluminum alloys are utilized here because of their lower density, corrosion resistance, and high thermal conductivity. Typically, three layers of aluminum are used in a brazing process. Most often the layers consist of variations of the aluminum association materials listed in Table 5.4 as well as a third layer consisting of nearly all aluminum with modifications to promote corrosion resistance.

In this type of system, silicon is leached out by manganese in the 3003 layer, leaving a corrosion-resistant third layer. The process described is critical, especially with automobile condenser production. In the case of condensers, the tubes used to carry high-pressure refrigerant need to be extruded. These extruded tubes need to be brazed in order to maintain the structural integrity of the thin design (100 μm).

TABLE 5.4
Typical Heat Exchanger Material Designations

	Designation	Si	Fe	Cu	Mn	Zn	Al
Standard layer alloy	3003	<0.6	<0.7	.05–.20	1.0–1.5	<0.1	Balance
Clad layer alloy	4343	6.8–8.2	<0.8	<0.25	<0.10	<0.20	Balance

Brazing is a process where nonferrous alloys are heated to a braze temperature of approximately 1100°F. A eutectic mixture will be observed in which bonding of the two metals occurs at a point that is below the individual melting points. This creates a sandwich of the three layers of material, which are linked to one another by their metallurgy. Typically, a brazed joint layer is around 40% of the strength of a base alloy (usually 60–75 MPa). The molten metal and a material applied to the part are called flux, which interacts with the aforementioned clad layer and flows to form a very strong joint upon cooling. Flux is a material that cleans the metal and allows for a smooth capillary transition of the clad layer. Fluxes for aluminum are typically compounds such as hexafluoro potassium silicate (K_2SiF_6) with some type of dispersant such as polybutylene.

Several types of brazing are used in the automotive industry. The main two are controlled atmosphere braze (CAB) and vacuum brazing. In CAB, a nitrogen atmosphere is provided as the part goes through a braze oven at a specific braze profile. The alloy plate or cladding is heated to flow, and fillets are formed. In Figure 5.7, an aluminum–silicone phase diagram shows the alloys that are important in automotive heat exchangers for controlled atmosphere brazing.

The furnaces used in CAB brazing contain a thermal degreaser, flux unit, and drying oven. The controlled atmosphere braze has some advantages over vacuum brazing. The dimensional requirements for a CAB braze are less demanding than those of vacuum brazing. The flux used in CAB brazing is noncorrosive and no post-process cleaning is necessary. In addition, the braze process is done on a continuous process versus the batch processing of vacuum brazing. Most importantly, the CAB process is far less capital intensive than the vacuum braze process. A typical aluminum material cross section is shown in Figure 9.4 in Chapter 9. A CAB system

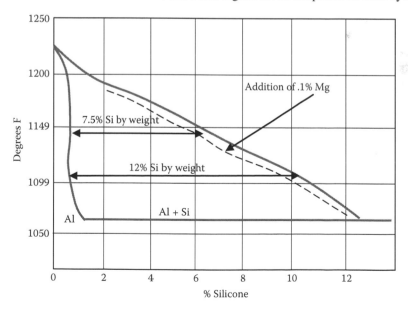

FIGURE 5.7   Aluminum–silicone phase diagram.

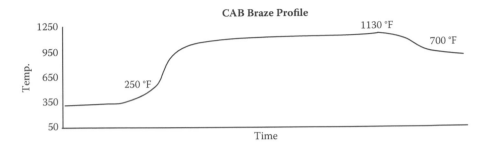

FIGURE 5.8   Controlled atmosphere braze profile.

includes a washer unit or thermal degreaser, a flux application unit, a dry-off oven, and a furnace. A profile is shown in Figure 5.8.

In the profile in the figure, the 250°F point is where the thermal degreasing takes place. The intent is to thermally remove oils utilized in earlier operations. The actual brazement takes place at the 1100°F point. It is necessary to keep a steady, consistent mass of material (thermal mass) going through the oven to maintain proper mass transfer of heat into the parts being brazed.

In vacuum brazing, a vacuum purge oven is used in a batch process. The aluminum parts, which can vary in size depending on the manufacture- and application-specific requirements ordered by the manufacturer, are placed in the oven. This process consists of a heating system, vacuum pumping system, nitrogen pumping system, and atmospheric controls. The unit is vacuum purged and then nitrogen is back-purged into the system. The furnace is quickly heated to the brazing temperature and soaked. Vacuum brazing is capital intensive and has a great energy cost. It is chosen because it can be loaded at room temperature; when the cycle is completed, the parts can be immediately processed because they are at room temperature.

The aluminum compositions used (3003 and 4343) are cast as ingots and homogenized. The 4343 alloys are hot-rolled to a thickness or approximately 2 mm; the 3003 layer is about 16 mm. The 4343 layer (or whatever clad layer is used) is then cold-rolled to an intermediate thickness for annealing. Finally, the material is cold-rolled to a thickness of about 100 μm, depending on the annealing process. The tempering process used is typically an O-temper or an H-14 temper. The H-14 is the cold-rolled process (which is mechanically soft) work hardened with strong anisotropy (directional dependency). H-14s have memory, thereby making them ideal for use in applications, such as automotive condensing, that are subject to stone damage due to their position in vehicles. When a stone strikes a condenser, it could perforate the condenser tube, thereby creating a leak. A tube tempered with H-14 versus an annealed O-temper without memory and having weak anisotropy will stand up to this type of damage more effectively.

5.5   PEM MANUFACTURING

As discussed in Chapter 10, proton exchange membranes used in the automotive industry are sulfonated tetrafluoroethylene copolymers (Figure 5.9) discovered by

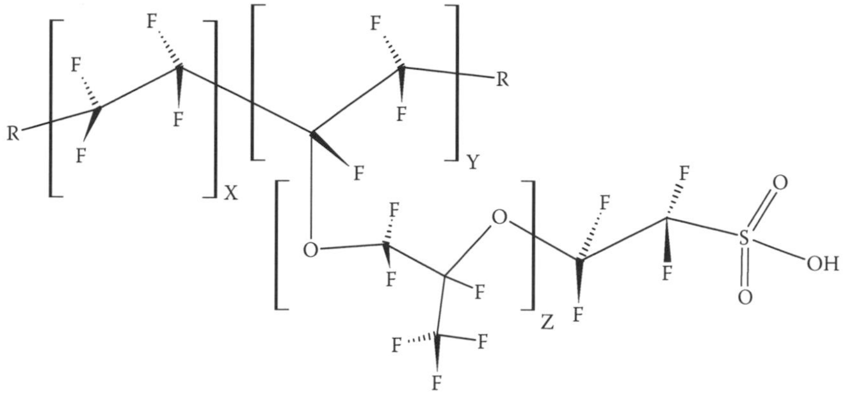

FIGURE 5.9   Sulfonated tetrafluoroethylene copolymer structure.

Walther Grot at DuPont in the late 1960s [6]. The material was called Nafion and was the first ionomer or synthetic ionic polymer. Figure 5.9 shows the polytetrafluoroethylene (Teflon) backbone supporting a perfluorovinyl ether group that is terminated by a sulfonate group. The thermal stability provided by the backbone from the Teflon makes the proton conductivity of the compound highly desirable. It is believed that the protons from the sulfonic acid group (SO_3H) can travel from one site to another. These membranes can allow movement of cations but not anions.

These compounds are synthesized by the copolymerization of tetrafluoroethylene and alkyl vinyl ether with sulfonyl acid fluoride. Preparation of sulfonyl acid fluoride takes place by pyrolysis of the respective oxide to give the olenfinated structure [7].

The thermoplastic produced is extruded into a film. The sulfonyl fluoride ($-SO_2F$) group present in the thermoplastic is converted to sulfonate ($-SO_3^-Na^+$) with NaOH. This is called the neutral form of Nafion, and it is converted to the acid for and cast into a thin film by heating in alcohol at 250°C. At around \$650/m², the Nafion membranes are rather expensive when compared to hydrocarbon membranes, which are also used.

5.6   NANOTUBE MANUFACTURING

Nanotubes are discussed in Chapter 10. They are generally 5–100 μm in length and have a diameter of 5–100 nm. Carbon is found in various configurations, including diamonds, graphite, CNTs (carbon nanotubes), and fullerene. CNTs are long graphite sheets rolled into a cylinder. They can have a single cylindrical wall (single-walled CNT [SWCNT]) or multiple walls (MWCNT). Multiple-walled nanotubes may contain from 100 to 1,000 atomic layers [8]. The MWCNTs contain a greater number of defects and as a consequence are less conductive and not as strong as SWCNTs [8].

Carbon nanotubes are considered the smallest known man-made structure that is chemically inert as well as self-supporting [8]. CNTs are also electrically more conductive than copper and have stiffness equivalent to that of diamonds. The most utilized manufacturing processes are listed in Table 5.5.

TABLE 5.5

Nanotube Manufacturing Process versus Yield

Nanotube Manufacturing Process	Yield	Process Notes
Laser ablation	<1.5 g	Metal catalyst needed
Electric arc	<1.5 g	Metal catalyst/pressurized chamber needed
Chemical vapor deposition (CVD)	>1.5 g	Patterned substrate used
HiPco process	>1.5 g	Pressurized CO_2 needed/high-temperature process

Each of the manufacturing processes listed in the table requires expensive equipment or the use of an expensive metal catalyst. The laser ablation and electric arc techniques will give single-walled carbon nanotubes in relatively small amounts (milligram to gram) in a matter of a few hours [8]. Another drawback is that the catalysts used in these techniques are difficult to remove from the finished product. The CNTs produced also require extensive purification as well as cleaning. The electric arc technique mentioned in Table 5.5 requires a closed pressurized vessel. Capital expense and energy cost as well as the potential for mishaps to occur are potential hazards with this process. The HiPco process is a very good process for mass production of carbon nanotubes. The process does require pressurization with a carbon monoxide atmosphere at very high temperatures along with a metal catalyst.

Because of the small size of nanotubes (<1 nm) and their excellent mechanical and electrical properties (depending on the hexagonal lattice and chirality), they have been recognized as ideal for nanocomposite structures. The relative tensile strength of theses structures can be as high as 200 GPa, with Young's moduli as high as 1 TPa.

A manufacturing technique for carbon nanotubes has been reported by Hui and Kin-Tak from the University of New Orleans and Hong Kong Polytechnic, respectively. They describe a process whereby an electrical current is induced through a carbon anode and a carbon cathode in which a nanotube is produced. The electrical current is produced from an arc welding power source. In this process, an exhaust hood is placed over the anode and a pressure chamber is not required. In addition to the elimination of the pressure chamber, a metal catalyst is no longer needed. This process produces SWCNTs in an effective manner.

These scientists suggest that the cathode and anode could be composed of graphite, activated carbon, and/or mixtures of the forms of carbon. The carbon cathodes are typically larger than the anodes, possibly with a bore in the center. The cathode may also be immersed in a tank of water for cooling. When the current is induced by the arc welding source, an inert atmosphere such as N_2 or H_2 is used. When the scientists induced the electric current through the anode, the anode was vaporized and a deposit formed on the surface of the cathode. The current was allowed to consume the anode and the deposits formed on the cathode were collected and purified. Reaction time was about 3 minutes for 1 g of product. When CNTs are produced in this manner, the decomposition temperatures are around 650°C versus the typical 500°C temperature found in other methods.

REFERENCES

1. http://goliath.ecnext.com/coms2/gi_0199-6220853/Process-machinery-advancements-in-mixing.html
2. Leab, D. J. 1985. *The labor history reader,* 336. Champaign: University of Illinois Press.
3. Rodriguez, F. 1989. *Principles of polymer systems,* 3rd ed., 348, 389, 391. New York: Hemisphere Publishing.
4. Bush, C. H. 2006. Electrode power. CNC West. http://www.cnc-west.com
5. Chenier, P. J. 1992. *Survey of industrial chemistry,* 322, 324. New York: VCH.
6. Connolly, D. J., and W. F. Longwood Gresham. 1966. Fluorocarbon vinyl ether polymers. U.S. Patent 3,282,875.
7. Heitner-Wirguin, C. 1996. Recent advances in perfluorinated ionomer membranes: Structure, properties and applications. *Journal of Membrane Science* 120:1–33.
8. Hui, D., and A. Kin-Tak Lau. 2002. *Nanotube composites.* Hung Hom, Kowloon: Hong Kong Polytechnic University.

6 Engineering Polymers, High-Temperature and -Pressure Applications, and Structural Polymers

6.1 INTRODUCTION

Engineering polymers for high-temperature and -pressure applications are in use extensively in original equipment manufacturers around the world. The high-strength properties of these materials are necessary to obtain the desired mass reductions, efficiency, cost, and performance required in today's market. In current automotive applications, several polymers predominate for thermoplastic seals, thrust washers, intake manifolds, and other components.

Several factors must be considered in seal design for rotating pressurized applications versus static sealing. Among these are processing factors such as injection or compression molding, physical properties such as tensile and yield strength, thermal properties, friction coefficient, and wear resistance. Rigid structural components such as fascias, grills, headlamps, and body panels have somewhat different requirements than the other components discussed in this book. We will discuss them in brief detail here.

6.2 DYNAMIC SEALING

As the automotive industry attempts to improve vehicle performance by reducing noise and vibration, decreasing weight, and increasing fuel performance, every aspect of design is considered. One would think that such a simple thing as a 10-g seal would not be worthy of scrutiny now that mass savings imperatives are measured in grams. Historically, cast iron seals were used in internal high-wear, high-temperature applications [1].

6.3 NEEDED PROPERTIES

An automotive seal, for instance, must be efficient; that is, it must have the ability to expand against a sealing surface quickly. This efficiency is related to the elastic modulus of the material, which is related to its stiffness [1]. The elastic modulus is the ratio of stress to corresponding strain in a material below the proportional limit on a

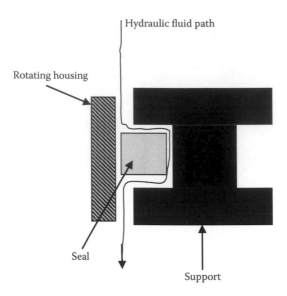

FIGURE 6.1   Diagram of rotating shaft seal.

stress–strain curve [2]. If the material is stiffer, it will require more energy to expand it against a sealing surface. Figure 6.1 shows a typical scenario from a transmission application with a rotating clutch housing, a polymer seal such as a polyetheretherketone or polyimide, and a solid support such as a sprocket support.

The difficulty in applications such as this is that the material chosen has to exhibit a variety of performance characteristics associated with component functionality—properties such as

- low coefficient of friction to reduce the drag against the mating component;
- good wear resistance against the mating contact surfaces;
- good dimensional stability and thermal stability; and
- high impact strength.

These types of rotating or dynamic seals must in a sense operate as valves, allowing some fluid to get through the seal when needed (to lubricate bearings) but being able to seal enough for a gear to be changed. Engineers and chemists must work together to get just the right design as well as work with raw material suppliers to get the right formulation to produce the optimum modulus. That is, a joint that opens and closes must be designed to comprehend all of the requirements for that seal, including thickness and material considerations. The chemist must tweak the filler content, cross-link density, grain direction, etc., to fulfill the testing requirements mentioned later.

6.4 AUTOMOTIVE REQUIREMENTS

When thermoplastic dynamic seals are designed for automotive applications, perhaps the most important parameter is the PV (pressure and velocity) value. PV is a measure of the seal's ability to withstand high pressures and velocities. A seal is placed in an apparatus that increases the pressure and velocity of the forces working against the seal until it fails [1]. This PV value at failure is directly proportional to the life of the seal. The units of PV are pounds per square inch and feet per minute. P is the hydraulic oil pressure in pounds per square inch, and V is the velocity of the rotating member in revolutions per minute. Table 6.1 shows the conditions that can be expected to be experienced by a dynamic seal in the field for today's vehicles. A minimum PV value of 500,000 is required for the engineering polymer to be utilized in a design.

The coefficient of thermal expansion must also be accounted for in design calculations. The normal operating temperature for transmission applications is 120°C. A material must be able to operate at this temperature and not expand to a greater volume than the mating parts allow. The coefficient of friction (μ) of the material should be considered as well when designing a dynamic seal. Rotating shafts require a lower value for μ in order to reduce wear.

Friction heat is generated when the two surfaces make contact. The higher the rubbing speed and compression load are, the more heat is generated. In a nonlubricated condition, the friction heat generated cannot be dissipated effectively; this results in a temperature increase at the rubbing interface. When the interface temperature is high enough to reach the glass transition temperature (T_g) or even the melting temperature (T_m) of one of the mating components, severe deformation occurs.

Production of automobiles is in the millions of units per year, so cost is a severe factor. A compression molded polyimide, for instance, is typically on the order of $0.33 per seal. However, this cost becomes insignificant given that these dynamic seals are located deep within the interior of the transmission, and thus labor for disassembly and time will make the cost of a failure quite severe.

When an automotive chemist chooses a polymer for dynamic sealing, he or she will have to consider the conditions present in Table 6.1. The two main polymers utilized today for these applications are the thermoset polyimides and polyetheretherketones. Polytetrafluoroethylene is also used for less severe applications. Wear test data are collected on an apparatus in which three circular pins of the polymer being examined

TABLE 6.1
Conditions for Rotating Shaft Seals

Rotating speed	6,500–7,000 rpm
PV value	300,000–1,000,000 psi*ft/min
Pressure in the radial direction	1.38 MPa (200 psi)
Differential line pressure	250–300 psi
Typical seal diameter	50–70 mm
Typical cross section	2.4-mm radius direction; 2.5-mm axial direction
Operating temperature	−40 to >150°C
Lube	Transmission fluid
Mating component	1010 steel (RB 80–85)

are mated against a washer made of AISI C-1018 steel. The sample is placed under a load while it is rotated at a speed from 10 to 5,000 rpm. During rotation, the friction force can be measured by a digital force gauge or collected using a load cell and data acquisition system. A thermocouple is placed inside the stationary steel washer to measure temperature. The test operates in a series of steps that will increase load, and the rotary spindle is driven by a variable speed direct current motor. The temperature and coefficient of friction are recorded manually several times during each step.

Test sequences start with the lowest pressure velocity value of 50,000 psi-ft/min and complete at the maximum pressure velocity value of 500,000 psi-ft/min. Two sequences are run. Sequence A is run at high pressure and low speed, and sequence B is run at low pressure and high speed. Tables 6.2 and 6.3 show sequence A and sequence B. The pressure and speed columns on the right multiplied by the area of the polymer being tested yield the PV value listed in that column. For test 1, for example,

$$(189 \text{ rpm})*(1075 \text{ psi})*(2)*(3.141)*(0.0392\text{in}^2 : \text{polymer area}) = 50,000 \text{ psi-ft/min}.$$

The coefficient of friction at each PV step is continuously monitored. Because the interface temperature keeps changing during the test at each PV step, the coefficients of friction therefore are recorded during the wear test. The values of the coefficient of friction tend to decrease as the steps increase [1].

In Table 6.4 (high-speed, low-pressure combination), the μ values for a polyimide dynamic seal for sequence B are shown. The lower the μ values are, the better is the wear performance of the material to its mating surface. At a given PV condition, the wear of a plastic material caused by increasing sliding speed is more severe than that caused by increasing pressure. Therefore, sequence B is more critical for

TABLE 6.2
Test Sequence A (High Pressure, Low Speed)

Test Step	PV (psi-ft/min)	P (psi)	V (shaft rpm)
1	50,000	1075	189
2	75,000	1313	232
3	100,000	1521	268
4	125,000	1696	300
5	150,000	1862	328
6	175,000	2006	355
7	200,000	2150	379
8	250,000	2404	423
9	300,000	2634	464
10	350,000	2845	501
11	400,000	3041	536
12	450,000	3225	568
13	500,000	3400	599

TABLE 6.3
Test Sequence B (Low Pressure, High Speed)

Test Step	PV (psi-ft/min)	P (psi)	V (shaft rpm)
1	50,000	129	1581
2	75,000	158	1930
3	100,000	182	2236
4	125,000	204	2491
5	150,000	223	2739
6	175,000	241	2952
7	200,000	257	3162
8	250,000	287	3536
9	300,000	315	3873
10	350,000	340	4183
11	400,000	364	4472
12	450,000	386	4743
13	500,000	407	5000

TABLE 6.4
Coefficient of Friction at Test Step for Polyimide

Test Step	PV (psi-ft/min)	μ
1	50,000	0.27
2	75,000	0.23
3	100,000	0.20
4	125,000	0.19
5	150,000	0.18
6	175,000	0.17
7	200,000	0.16
8	250,000	0.14
9	300,000	0.13
10	350,000	N/A
11	400,000	N/A
12	450,000	N/A
13	500,000	N/A

TABLE 6.5
Coefficient of Friction at Test Step
for Polyetheretherketones

Test Step	PV (psi-ft/min)	μ
1	50,000	0.32
2	75,000	0.36
3	100,000	0.29
4	125,000	0.25
5	150,000	0.25
6	175,000	0.23
7	200,000	0.18
8	250,000	0.16
9	300,000	N/A
10	350,000	N/A
11	400,000	N/A
12	450,000	N/A
13	500,000	N/A

dynamic sealing applications. In Table 6.5 the same data are shown for the thermo-plastic polyetheretherketone.

Thermal capability (T_g and T_m) and the heat dissipation capability at the interface determine the wear performance of a plastic material. Thermal capability of a plastic material depends on its chemical structure and material composition. The major factors that contribute to thermal stability in polymers include [3]:

- aromatic and heteroaromatic groups in a structure;
- increased crystallinity in a polymer;
- polar side groups; and
- intermolecular cross-linking.

Two of the examples listed in this chapter (polyimide and polyetheretherketone) contain several aromatic groups in their structure.

The heat dissipation capability is affected by both internal and external lubrication in these applications. The effect of this external lubrication is usually more significant as it removes heat. If the interfaces are always covered with a homogeneous layer of lubricant, the wear performance of two plastic materials with different thermal capabilities may become the same, even though significant differences are expected if running occurs in dry conditions [1].

6.5 MATERIALS AND PROCESSING

As mentioned previously, two polymer materials are used today to replace the older cast iron seals previously used for dynamic high-temperature and high-pressure applications such as transmission shaft seals. The polyimides and polyetheretherketones have high wear resistance and can be used at high operating temperatures (up to 287°C) [4]. Economically, however, there are cost penalties for using a polyimide material versus a polyetheretherketone. The polyimide must be compression molded and subsequently machined because it is a thermoset material. The polyetheretherketones are injection molded, which is better from the cost-imperative point of view. The advantage of injection molding a polyetheretherketone is that no machining is required after the part is made. The material is injected into a mold, packed, and cooled and no further processing is required. A machining operation can add as much as $0.33 to a seal's cost.

The compression molded polyimide comes in a powdery brittle state as it is added into the compression mold's cavity. Pressure is applied and at the processing temperature cross-linking takes place. Due to the thermal cycling of the mold, the process is not very efficient and molds are subject to a shorter life [5]. All of these factors add to the cost of the part.

6.6 THERMAL PROPERTIES

At the glass transition temperature (T_g), a thermoplastic material changes from a glassy state to a rubbery state. The properties of the material also change significantly. T_g values most often listed for polymers correspond to stiffening temperatures [3]. The coefficient of thermal expansion usually doubles below T_g for these materials. Materials above the T_g may be functional, but the performance may become unpredictable because most thermoplastic components are designed based on properties tested below T_g.

The melting point, T_m, is the transition temperature at which a thermoplastic material changes from a rubbery state to a fluid state. Above T_m, a thermoplastic material completely loses its function. For a thermoplastic component, the optimum application temperature is the usage range below its T_g.

Thermoset materials such as polyimides behave quite differently from thermoplastic materials. They usually do not exhibit noticeable temperature transitions. The term *thermoset* is applied to materials that, once heated, react irreversibly so that subsequent applications of heat and pressure do not cause them to soften and flow [3]. This property leads to good thermostability.

6.7 FILLERS

A polymer's properties control the characteristics of the part made from it; however, the polymer is only one of several constituents. Polymers are compounded with an array of other materials. The polyimides and polyetheretherketones discussed here use reinforcing fillers and curing agents. Curing agents are used to form the network of cross-links that guarantee elasticity rather than flow [3].

Unlike inert fillers that have little effect on physical properties, reinforcing fillers such as carbon black and Teflon add to the mechanical properties of the polymer [6]. Both of the polymers for rotating dynamic seals examined here contain carbon black as a reinforcing filler. Teflon is added to polyetheretherketone to increase the lubricity properties of the polymer. Ninety percent of carbon black is obtained through furnace black, which involves the combustion of thermal cracking of hydrocarbons [7]. This reaction takes place at 1,200–1,400°C:

$$-CH_{2-} \rightarrow C + H_2$$

$$-CH_{2-} + 1\tfrac{1}{2} O_2 \rightarrow CO_2 + H_2O \tag{6.1}$$

Another benefit of carbon black may be the potential for protection against oxidation. Due to the higher temperatures over an extended period of time, thermal oxidation can occur in a polymer. The mechanism for thermal oxidation is given in Figure 6.2.

There are three states in thermal oxidation: initiation, propagation, and oxidative chain branching [8]. Chain-breaking antioxidants are effective in the propagation

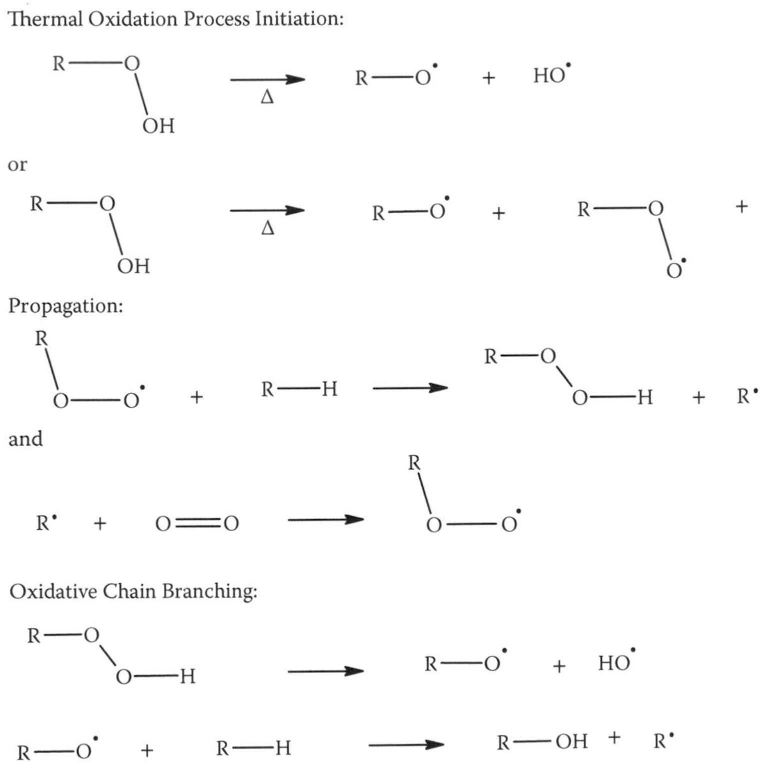

FIGURE 6.2 Thermal oxidation process.

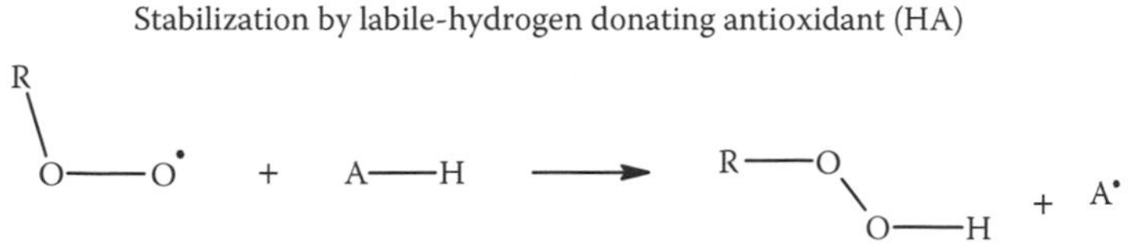

Stabilization by labile-hydrogen donating antioxidant (HA)

FIGURE 6.3 Stabilization by labile hydrogen donating antioxidant.

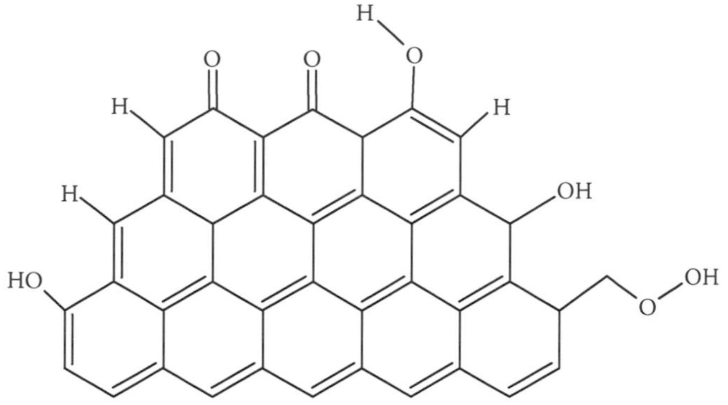

FIGURE 6.4 Structure of carbon black.

stage of this process. They react with the radicals that propagate oxidation, thus reducing the hydrogen donating antioxidant (HA) shown in Figure 6.3.

The edge of the structure of carbon black is given in Figure 6.4 [8]. Carbon radicals can migrate along a polymer chain for significant distances [8]. With a carbon black filler, there is the possibility that radicals will migrate along the chain. Commercial carbon blacks have between 3 and 4% oxygen combined on the surface and much of this is believed to be present as phenolic groups. Carbon black can thus be regarded as a giant phenol and as such should be capable of functioning as a labile-hydrogen donor [8].

6.8 POLYETHERETHERKETONES

The most widespread polymer currently in use for high-speed dynamic sealing applications in automotives is polyetheretherketone. These thermoplastic aromatic polymers are used in the aerospace, electronics, and nuclear industries [9]. These materials have excellent mechanical- and chemical-resistance properties, which permit polyetheretherketones to be used in many engineering applications—often in harsh environments [10]. These polyetheretherketones are the product of the reaction of 1,4-dihydroxybenzene and 4,4′-difluorobenzophenone [8]. Figure 6.5 shows this reaction.

Various amounts of filler contents are available; typically, 33% glass fiber is used. T_g of 143°C and melting temperature of 342°C, with a continuous use temperature of 260°C, are observed without thermal degradation [11]. This material dissolves in concentrated sulfuric acid and is not chemically attacked by water or pressurized steam. Most importantly, it can be injection molded and is capable of holding

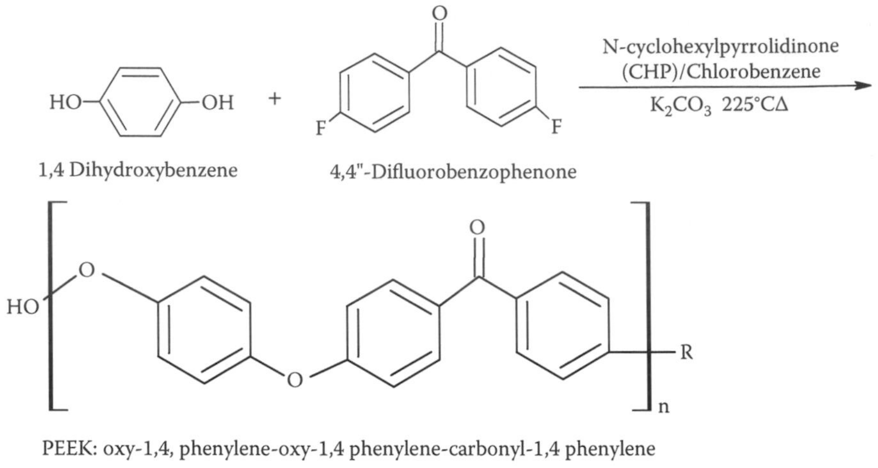

PEEK: oxy-1,4, phenylene-oxy-1,4 phenylene-carbonyl-1,4 phenylene

FIGURE 6.5   PEEK production.

tight tolerances that eliminate machining. By thinking in this manner, the chemist provides insight into the manufacturing process and invaluable help to the overall design and function of the vehicle.

6.9   POLYIMIDES

Another class of material that is currently in use is polyimides. They are generally the reaction products of an aromatic diamine with a dianhydride such as pyromellitic dianhydride (shown in Figure 6.6) [3]. They are also synthesized by reacting 1,4-phenylene diisocyanate with pyromellitic dianhydride [12]. Aromatic heterocyclic polyimides exhibit outstanding mechanical properties and excellent thermal and oxidative stability. These compounds are of major industrial importance [13]. The aromatic ring structure in these polymers induces rigidity and high melting points. This material is a thermoset and therefore no plastic flow is observed at higher temperatures, but rather thermal degradation is observed. This material cannot be melt-processed and has no T_g, but it is compression molded [5]. After compression molding, it must then be machined into its final form. These machining applications can contribute as much as 30% to the part's cost.

6.10   POLY(TETRAFLUOROETHYLENE)

Another dynamic seal type used is the poly(tetrafluoroethylene) (PTFE) seal utilized on applications with PV values that are not as stringent as those of clutch systems. PTFE is a fluoropolymer that has many other applications besides automotive. DuPont has the trade name Teflon for this material. Other usages include nonstick cookware, lubricants, bearings, bushings, gears, and plumbing materials. It is useful as a resistant material to corrosive and reactive chemicals. It is an extremely nonreactive material that was discovered by accident by Roy Plunkett when he

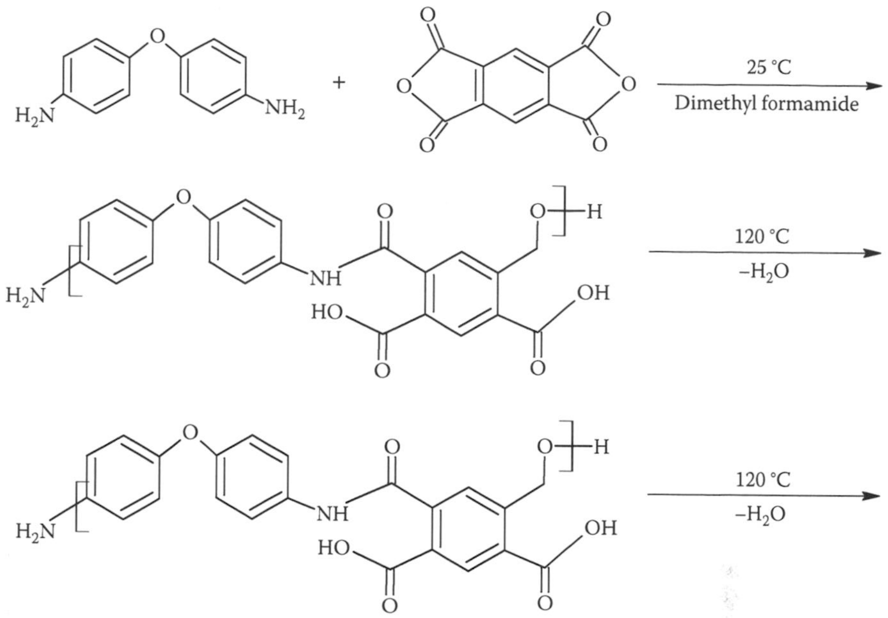

FIGURE 6.6 Polyimide production.

was attempting to work out a new refrigerant [14]. PTFE can be synthesized by the polymerization of tetrafluoroethylene (shown in Figure 6.7).

The emulsion polymerization is conducted under pressure using free radical catalysts. It may also be manufactured by direct substitution of hydrogen on polyethylene with fluorine. The emulsion polymerization can produce high molecular weight polymers at a fast rate. This is a type of polymerization that incorporates water, monomer, and a surfactant. Polymerization takes place in latex particles typically around 100 nm in size. PTFE melts around 327°C, begins to degrade at 260°C, and has a density of 2.2 g/cc [15]. The coefficient of friction is <0.1. This extremely low value allows the material to be ideal for dynamic rotating seals at high pressures, temperatures, and wear.

The material has some drawbacks because the cross-linking is limited due to its nonreactivity. PTFE tends to take a compression set and is subject to creep. If a tight tolerance dynamic sealing application is needed, then the creep can take the form of the mating components and may not essentially be a bad thing if the filler material is chosen carefully. PTFE particles are often used as a filler. For instance, a

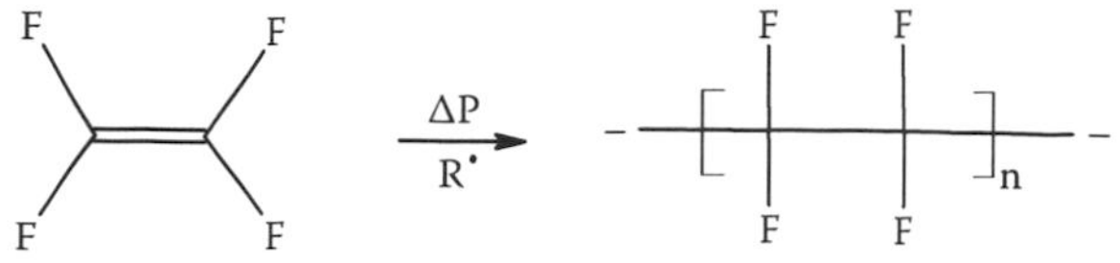

FIGURE 6.7 PTFE synthesis reaction.

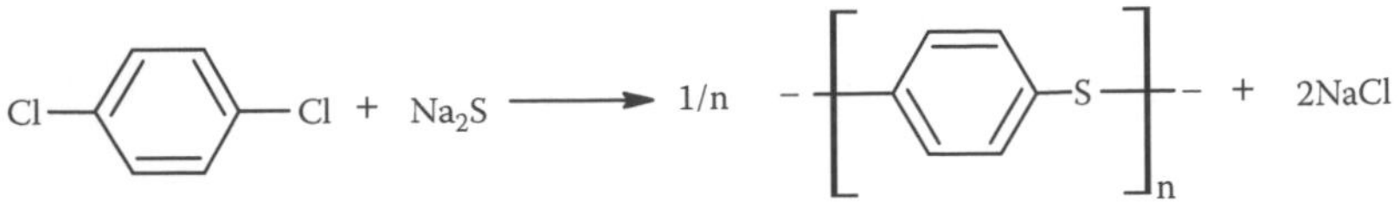

FIGURE 6.8 Polyphenylene sulfide synthesis reaction.

very effective polyetheretherketone seal utilizes a Teflon or PTFE filler. Embedded lubricants such as molybdenum disulfide and mineral oil have also been used to improve the performance of these dynamic seals.

6.11 PPS

Another engineering polymer that is utilized in automotive applications is PPS—polyphenylene sulfide, which is a sulfide-linked polymer shown in Figure 6.8. The material is manufactured by reacting dichlorobenzene with a sulfide. A commercial production process for PPS was developed by Phillips Petroleum and designated Ryton [16]. The aromatic ring gives this polymer excellent thermal resistance and dimensional stability. In addition, it is resistant to alkalis and acids. In automotive applications, the main usages are in transmissions as thrust washers and accumulator pistons. In thrust washers, a load is applied to the washer, which is mounted along a shaft to prevent movement along that axis. Accumulator pistons are spring-loaded pistons that move up and down in a bore in response to fluid flow on either side of the piston, thus absorbing shocking shifts. In both of these applications, tight tolerances, low CTEs (coefficients of thermal expansion), and high mechanical strength are required. Polyphenylene sulfide is another example of design that the automotive chemist can provide direction.

REFERENCES

1. Phlegm, H. K. 2001. A comparison of molded polyaryletherketones and compression molded polyimide polymers in automotive applications. Master's thesis, University of Detroit-Mercy, pp. 1, 7, 11.
2. Potter, M. 1990. *Fundamentals of engineering,* 3rd ed., 10–12. Okemos, MI: Great Lakes Press.
3. Rodriguez, F. *Principles of polymer systems,* 3rd ed., 22–23, 44–45, 296–301, 503–504. New York: Hemisphere Publishing Corporation.
4. *GE plastics design guide.* 1999. p. 10.7. Pittsfield, MA: GE Plastics.
5. Properties of DuPont Vespel parts. 1993. pp. 6–7, 10. Wilmington, DE: DuPont Corporation.
6. Billmeyer, F. 1962. *Textbook of polymer science,* 1st ed., 528. New York: John Wiley & Sons Publishing.
7. Chenier, P. 1992. *Survey of industrial chemistry,* 2nd ed., 112. New York: VCH Publishers Inc.
8. Hawkins, W. 1984. *Polymer degradation and stabilization,* 1st ed., 59–60. Berlin: Springer–Verlag Publishing.

9. Johnson, R. N., A. G. Karnham, F. A. Clendinning, W. F. Hale, and C. N. Merriman. 1967. Facile synthesis and properties of semicrystallization co-poly (arglene ether ketone) containing hydroquinone and photoalazinone. *Journal of Polymer Science, Part A* 5:2375.

10. Cogswell, F. N. 1992. *Thermoplastic aromatic polymer composites.* Oxford, England: Butterworth–Heinemann.

11. *Victrex PEEK properties guide.* 1997. pp. 6–7. Thornton Cleveleys, UK: Victrex Technology Center.

12. Barikani, M., and S. Mehdipour. 1999. Synthesis, characterization and thermal properties of novel arglene sulfure ether polyinides and polyamides. *Journal of Polymer Science, Part A* 37:2245.

13. King, F. 1985. Synthesis and characterization of new soluble aromatic polyimides. *Journal of Engineering Thermoplastics.*315.

14. Plunkett J. 2008. Chemical Heritage Foundation. Retrieved July 20, 2008.

15. http://www2.dupont.com/Teflon_industrial/en_US

16. Wayne H., and J. Edmonds. 1967. Philips Petroleum Company Research Center (Bartlesville, OK). Patent 3,354,129, 1963. Issued November 21, 1967.

7 Power Train Applications

7.1 INTRODUCTION

The chemist's role is not limited to materials and analysis of designed components. The role also extends into power train applications and processes. "Power train," refers to the set of components that generate power (engine) and deliver power (transmission, differential, and drive shafts) to the road. The combustion process, lubrication of components, and cooling requirements—along with ways to improve these processes and requirements—must be understood. The automotive chemist's tools and training allow for these challenges to be met properly as we head into the future.

7.2 FUEL COMBUSTION

The combustion and airflow system in an automobile delivers a fresh charge of air and fuel into the combustion chamber and then converts the chemical energy into mechanical energy by the combustion process. Equation 7.1 shows how the process works:

$$C_3H_8 + [5O_2 + N_2](air) \rightarrow 3CO_2 + 4H_2O + N_2 + Heat \qquad (7.1)$$

It is very important for the power train designers to understand this process because the combustion airflow system influences efficiency, emissions, performance, heat rejection, structural requirements, and noise. Figure 7.1 shows a simple example of the combustion airflow system for a conventional spark injection (SI) engine. The air and fuel are mixed together in the intake system, inducted through the intake valve into the cylinder, and mixed with residual gas. The mixture is then compressed and combustion is initiated toward the end of the compression stroke by a spark discharge. A turbulent flame is initiated at spark discharge. The combustion process can be thought of as a controlled burning process rather than an explosion. Figure 7.2 shows the cylinder pressure versus crank angle for the four-stroke engine used in automobiles.

The crank angle in degrees expresses the minimum spark advance for best torque (MBT). Maximum engine torque is expressed where the spark timing is at the MBT point. Likewise, the specific fuel consumption will be at a minimum at the MBT point.

The optimum air–fuel ratio can be expressed as the stoichiometric or theoretical combustion ratio. This is the mixture in which fuel will be burned completely. In other words, all the carbon will be converted to carbon dioxide, all hydrogen converted to water, and all sulfur converted to sulfur dioxide. If there are unburned components in the exhaust gas, such as C, H_2, or CO, the combustion process is incomplete. Carbon monoxide is produced if there is insufficient oxygen to oxidize the fuel fully. Nitrogen

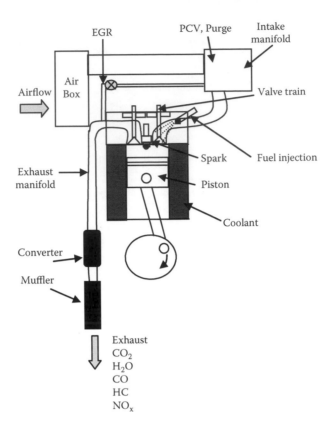

FIGURE 7.1   Diagram of internal combustion engine.

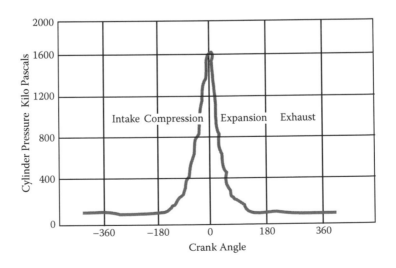

FIGURE 7.2   Crank angle versus cylinder pressure.

oxides are produced at higher temperatures in the presence of oxygen. Nitrogen molecules will react and form NO and NO_2. For stoichiometric combustions,

$$C_aH_b + (a + b/4)(O_2 + 3.773N_2) \rightarrow aCO_2 + \left(\frac{b}{2}\right)H_2O + 3.773\left(a + \frac{b}{4}\right)N_2 \quad (7.2)$$

With $y = b/a$,

$$CH_y + \left(1 + \frac{y}{4}\right)(O_2 + 3.773N_2) \rightarrow CO_2 + \left(\frac{y}{2}\right)H_2O + 3.773\left(1 + \frac{y}{4}\right)N_2 \quad (7.3)$$

Starting with the stoichiometric air–fuel ratio or the ideal fuel ratio, burning would consume all the fuel without leaving unburned constituents. If the air content is higher than this ratio, then the mixture is said to be lean; if the air content is lower than the stoichiometric ratio, then the mixture is rich. The air–fuel equivalence ratio (Φ) shown in Equation 7.4 is considered rich if the value is greater than 1:

$$\Phi = \frac{\left(A/F\right)_{Stoich}}{A/F_{Actual}} \quad (7.4)$$

Unburned fuel can contribute up to 1–2% of emissions on a modern engine due to incomplete combustion or design issues such as cylinder-to-cylinder variation and nonhomogenous air–fuel ratios. Combustion in oxygen is a radical chain reaction that will contain intermediates. It is thought to be initiated by the abstraction of a hydride from the fuel to oxygen. A hydroperoxide radical will react to give hydroperoxides and hydroxyl radicals. The intermediates cannot be isolated, but the nonradical intermediates are stable.

Temperature of the intake air will play a role in the combustion process. The temperature affects the air density, which will in turn have an impact on the mass of the air in the cylinder. For every 3°C increase in intake temperature there is an approximate 1% loss in density, which results in about a 1% loss in power. The effects of intake temperature on combustion are not limited to the power loss due to density. It also affects spark knock, spark retardation, elevated under-hood temperatures, and elevated exhaust valve temperatures.

In a combustion reaction, the fuel and oxygen from the air will result in a high amount of nitrogen. The nitrogen will be oxidized to various nitrogen oxides (NO_x). In addition, sulfur will produce sulfur dioxide and carbon will produce carbon dioxide. Because it is practically impossible to have complete combustion, the automotive chemist must devise ways of reducing the impact of the pollutants produced. This is demonstrated by the use of urea injection in diesel engines to minimize the effects. The design engineer must also devise ways of manufacturing and design to improve turbulent flow and efficiency. For instance, in aluminum engine blocks, a steel liner can be compressed by lowering its temperature in liquid nitrogen and then placed in the cylinder and allowed to expand for a press fit.

Design for turbulent combustion is another critical way that the design engineer can affect or improve the stoichiometry of a system. Turbulence helps the mixing process between the fuel and oxidizer. Injection ports, piston design, and cylinder design all contribute to good combustive characteristics. Further improvements in combustion are achieved by catalytic converters and exhaust gas recirculation.

7.3   DIESEL INJECTION (UREA INJECTION)

Another recent advancement involving chemists and design is the development of diesel emissions fluid (DEF) systems. Diesel emissions fluid is a 35% urea and 65% water mixture injected into the exhaust stream of a diesel vehicle to improve or reduce the NO_x emissions. The urea injection works by decomposition of urea into ammonia when injected into the exhaust stream:

$$H_2NCONH_2 \xrightarrow{\Delta} NH_4 + NCO \rightarrow NH_3 + HNCO \tag{7.5}$$

The ammonia then reacts with NO_x over a diesel oxidation catalyst (DOC) downstream of the injection point to reduce NO_x emissions:

$$4NO + 4NH_3 + O_2 \rightarrow 4N_2 + 6H_2O \tag{7.6}$$

$$2NO_2 + 4NH_3 + O_2 \rightarrow 3N_2 + 6H_2O \tag{7.7}$$

These DEF systems are generally mounted on the side of the frame of the vehicle. The fill point is generally under hood. The actual diesel emissions fluid is a 35% urea–water mixture. Because this mixture freezes at –11°C, a heated EPDM (ethylene–propylene diene monomer)/nylon 6,6 line is used to deliver the urea from the tank to the exhaust stream. A sensor controls the flow of the urea to obtain the NO_x reduction. This system can reduce NO_x emissions by up to 40%. The urea refill rate is about once every 3 months based on typical driving schedules.

7.4   ENGINE OIL

Engine oil is obviously another area of great importance for the automotive chemist, and the design and analysis of these oils play a tremendous role in the operation of a motor vehicle. Engine oils lubricate engine parts, inhibit corrosion of these parts, assist with vehicle cooling, and help seal some of the components in the vehicle. Traditional engine oil is derived from base petroleum stock and refined to meet the requirements of the American Petroleum Institute (API). The API sets minimum performance standards for a motor oil. The base stock will contain additives to improve the oil's detergency, extreme pressure performance, and corrosion inhibition. The API groups oils according to base stock, as shown in Table 7.1.

Some basic terms should be defined for engine oils:

TABLE 7.1
API Base Stock Oil Groups

Group	Oil type
I	Fractionally distilled petroleum followed by solvent extraction
II	Fractionally distilled petroleum followed by hydrocracking
III	Group II oils further hydrocracked to produce higher viscosity index values
IV	Base stock oil of polyalphaolefins (PAOs)
V	Polyol esters/polyalkylene glycols (PAGs)/others

Cracking: Heating of higher molecular weight petroleum fractions in the presence of a catalyst to give a lower molecular weight fraction [1].

Dewaxing process: Lubrication feedstocks typically have an increased wax content from other processing and refining (de-asphalting). These waxes can be removed catalytically or by solvent dewaxing. A reactor contains a dewaxing catalyst followed by a second reactor with a hydrogen finishing catalyst to saturate olefins from the dewaxing reaction [3].

Fractional distillation: Separation of a mixture of components or fractions by their boiling points by heating them to that point. This is a continuous process in the petroleum industry and the most common separation technique used in refineries (Figure 7.3) [2].

Hydrotreating: Generally, the first process before cracking. Petroleum fractions are reacted with hydrogen at 400°C with a cobalt oxide/molybdenum oxide catalyst [1]. This decreases the amount of nitrogen and sulfur compounds and prevents poisoning.

Kinematic viscosity: Viscosity is a fluid's internal resistance to flow, or its thickness. Kinematic viscosity is a measure of the ratio of the viscous force to the inertial force of the fluid or $v = \mu/\rho$, where μ is the dynamic viscosity (in centipoise [cP]), ρ is the density (in grams per cubic centimeter), and v is the kinematic viscosity (in centistokes [cSt]).

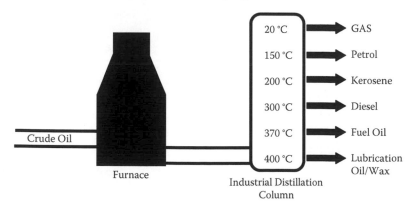

FIGURE 7.3   Fractional distillation of crude.

Reforming: Catalytic reforming is the process of increasing the number of double bonds on a petroleum product but maintaining the same number of carbon atoms. This process is done at high temperatures in the presence of a platinum or rhenium catalyst on alumina [1].

Solvent extraction: Solvent extraction is a method of separating compounds based on their solubility in different immiscible liquids. In industrial processes, solvents are typically transferred from an aqueous phase to an organic phase.

Viscosity index: Viscosity index is a measure of the change of kinematic viscosity with temperature and indicates an oil's ability to lubricate with a change in temperature. The VI scale set by the Society of Automotive Engineers (SAE) is 0 (worst) to 100 (best). VI = 0 is naphthenic oil and VI = 100 is paraffinic oil.

7.5   ENGINE OIL FUNCTION

Chemists play a role in the addition of additives to improve the properties of motor oil. As steel and aluminum parts move closely against one another, friction between the two parts creates heat and wears away the metallic surfaces. This will lead to decreased efficiency, increased fuel consumption, decreased power output, and eventually engine failure. Engine additives are devised to combat these consequences. A list of engine additives follows; however, rather than discussing each one in depth, we will discuss only some of the more important ones:

- extreme pressure additives;
- detergent additives;
- metal deactivators;
- corrosion inhibitors;
- antifoam agents;
- friction modifiers;
- viscosity index improvers;
- antioxidants;
- antiwear additives; and
- emulsifying agents.

Motor oil will create a film between the moving metallic parts, thereby lessening the contact between them, decreasing the amount of friction and wear, and removing heat. Toward the top of the piston around the compression rings, temperatures can reach up to 160°C, while in diesel engines the temperatures can rise by another 150°C. Secondary heat exchangers called engine oil coolers (EOCs) are a design feature added to minimize these effects. They are basically plate coolers made of four or five thin plates located on an external loop in the airflow of the vehicle.

Another function of the motor oil is to coat the parts to minimize exposure to oxygen. Corrosion inhibitors within the oil assist in this function. Sludge buildup is a result of degradation of the oil after exposure to the heat and extreme pressure conditions of the engine. Many motor oils have detergents and dispersants to combat this problem. However, microscopic particles will invariably be produced from wear and cause erosion. The oil filter is another design feature that will help with the overall process.

Engine oil lubricates rotating parts in the crankcase of a vehicle as it is applied to crankshaft journals, rods, and bearings. During testing and design of oil coolers, thermocouples are placed at the engine sump at the bottom of the crank to measure temperature of the oil. In many engines, turbulators are placed in the oil pan to spread the oil on the reciprocating members of the engine. The oil pump will send oil through the filter into the galleries and across critical parts such as the camshaft bearings. Oil from holes in the main journals will move through passageways inside the crankshaft to exit holes in the rod journals to lubricate the rod bearings and connecting rods. Passageways through the rods carry oil from the rod bearings to the rod–piston connections and lubricate the contacting surfaces between the piston rings. Interior surfaces of the cylinders are lubricated from oil coming from the rod bearings. There it will be a barrier between the piston rings and cylinder walls before oil returns to the sump.

7.6   ENGINE OIL GROUPS

Table 7.1 shows engine oil groups from the API. Within the five oil groups, group I is base stock from fractionally distilled petroleum with some further refinement. Group II is base stock as in group I, but it has been hydrocracked for purification. Group III base stock will have a higher viscosity index than group II due to further hydrocracking. Group IV oils contain polyalphaolefins and group V oils are a group used to catch any oil not described by one of the other groups. The synthetic oils discussed later are contained in this latter group.

7.7   ENGINE OIL GRADES

Engine oil is graded based on an SAE coding system based on kinematic viscosity. The grades range from 0 to 60 in increments of 10. An oil grade with a "W" behind it designates a winter or cold-start grade; this means that the oil will have a better flow viscosity at lower temperatures. For single-grade oils, viscosity is measured at 100°C and the range that the viscosity falls in is called its weight. A single-grade oil's viscosity will increase on a logarithmic scale as the temperature increases.

A specific motor oil will have a high viscosity when cold and a lower viscosity when hot. The viscosity difference in single-grade oils will be too large to provide adequate protection for today's vehicles. Viscosity improvers such as polybutene are added to decrease the difference between the two extremes. It is these additives that designate a multigrade oil, which will have the viscosity of the base number when cold and the viscosity of the second number when hot. Multigrade oils have two grade designations. For instance, the "30" in the 5W30 grade means that the viscosity of this oil at 100°C corresponds to the viscosity of a single-grade 30 oil at the same temperature. The first number, "5," is associated with the winter measurement and is not rated at a single temperature. A multigrade with a first number of 5 can be utilized as well as a single-grade 5. SAE procedures test a 5W oil at –30°C; a 0W oil is tested at –35°C and a 10W oil is tested at –25°C on a cold crank simulator.

7.8 SOME IMPORTANT ADDITIVES

As mentioned earlier, oil additives play a very important role (such as that of VI improvers for multigrade oils). Antiwear agents such as zinc dialkyldithiophosphate (ZDDP) are shown in Figure 7.4. These compounds feature zinc bound to the anion of dithiophosphoric acid. ZDDPs are soluble in mineral and synthetic oils. This antiwear additive is present in most commercial oils. Its quantity is limited in order to minimize interactions with catalytic converters. ZDDP also contains calcium and protects engine oil from oxidative breakdown and sludge formation. Concentrations in fluids are around 1–2% [4]. These compounds are manufactured by treating phosphorus pentasulfide with an alcohol. Zinc oxide is then reacted with the resulting dithiophosphate:

$$2 (RO)_2PS_2H + ZnO \rightarrow Zn[(S_2P(OR)_2]_2 + H_2O \qquad (7.8)$$

The compound is monomeric but it also exists with dimers and oligomers. These compounds are also used as antioxidants and corrosion inhibitors. Molybdenum disulfide is also reported to have antiwear capabilities as well as to function as a friction reduction agent.

Another excellent antiwear compound is tricresyl phosphate (TCP). These compounds are used as a high-temperature antiwear and extreme pressure additive. Tricresyl phosphate is an organophosphate made from cresol and phosphorus oxychloride (shown in Figure 7.5). These compounds come in the ortho-, meta-, and para-cresyl phosphates. They are also used as gasoline additives [4].

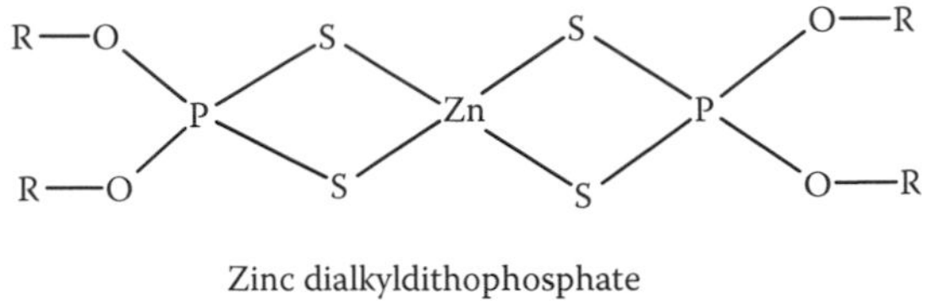

Zinc dialkyldithophosphate

FIGURE 7.4 Diagram of zinc dialkyldithiophospate.

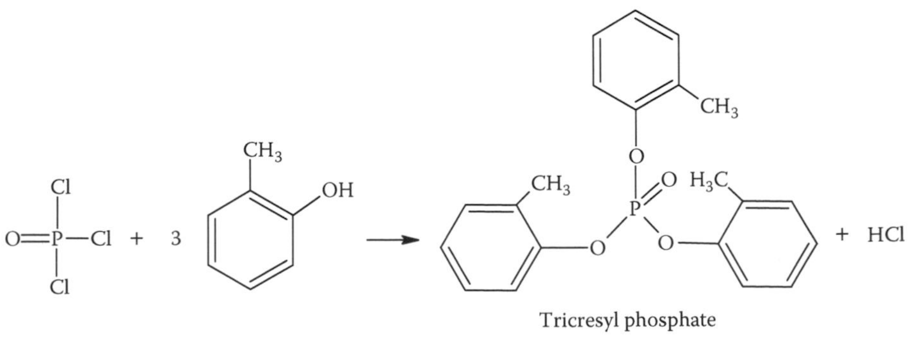

Tricresyl phosphate

FIGURE 7.5 Tricresyl phosphate production.

7.9 SYNTHETIC LUBRICANTS

In the late 1930s and early 1940s, German scientists began working on synthetic oils for military needs because pressure from the Allies caused a lack of crude oil. The scientists noticed that these products remained operable in the cold Russian winter during the war. This property aroused further interest and prompted further study into the products' structures and capabilities by the German as well as the Allies' scientists. Synthetics are made from group III mineral oils or group IV and V non-mineral-based oils. Some key true synthetics are the synthetic esters and polyalphaolefins (PAOs), diesters, polyolesters, and alkylated naphthalenes. Group II has been described as synthesized hydrocarbons. Group III base oils are considered synthetics in the United States only.

Part of the value of synthetic lubricants is the ability to have increased control over the process that manufactures them. Better control of processing translates to good mechanical properties at both temperature extremes. In Table 7.2, we can see that synthetics are substantially higher in viscosity index. The wider temperature range and lower pour points are due to their specially designed properties. Viscosity index improvers are not needed in the quantities, as they are in the group I, II, and III oils. As a result, when the oil ages, degradation will not be as quick or severe. The improved properties of a 0W synthetic oil can be seen in Table 7.3. Some manufactured oil change intervals can be as long as 15,000 miles. Manufacturers can use synthetic oil bases of PAOs, synthetic esters, polyalkylene glycols (PAGs), alkylated naphthalenes, and silicate esters.

TABLE 7.2
Viscosity Index of Various Oil Groups

Oil Group	Viscosity index
I	103–108
II	113–119
III	≥140
Synthetics (V)	187

TABLE 7.3
Properties of Synthetic Oil

SAE Grade	0W-40
cSt at 40°C	80
cSt at 100°C	14.3
Viscosity index	187
Pour point (°C)	−54
Flash point (°C)	236

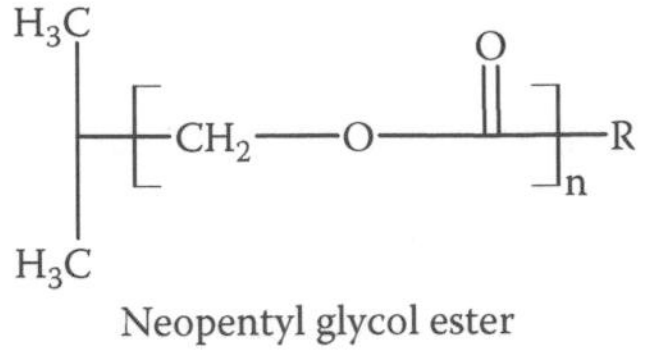

Neopentyl glycol ester

FIGURE 7.6 Neopentyl glycol ester.

7.10 SYNTHETIC ESTERS

Synthetic esters like polyol esters or neopentyl polyol esters are made from monobasic fatty acids and polyhedric alcohols with a neopentyl structure (see Figure 7.6). On these molecules, there are no hydrogen atoms on the beta carbon. This carbon is where thermal attack occurs on diesters and eliminating it improves the thermal stability of the molecule. Polyol esters have an increased number of ester groups versus diesters. This feature increases polarity, which will affect the lubricity of the oil at elevated temperatures and give it an advantage over PAOs.

7.11 POLYOLEFINS

Polyolefins—specifically, polyalpha olefins—are made from polymerizing alpha olefins where the double bond is located at the alpha carbon atom. Figure 7.7 shows 1-hexene. The double bond is located between the first and second carbon atoms and, as a result, they have flexible alkyl branching groups on every opposite carbon in the chain. The flexibility due to the different configurations makes crystallization difficult. Because polyalpha olefins are able to remain flexible at lower temperatures, they are ideal as a synthetic motor oil. Some advantages versus disadvantages of synthetics are shown in Table 7.4

7.12 AUTOMATIC TRANSMISSION FLUID (ATF)

The severity of the life of automatic transmission fluid is less severe than that of engine oil. Without the microscopic coke particles, soot, fuel interaction, and other contamination associated with motor oil, ATF must contend only with particles from friction plates and bearings from inside the transmission. Transmissions typically contain a nylon screen or filter of around 150 μ to filter out these particles; however, changing the fluid is the only way to eliminate them. The metallic particles present can contribute to the oxidation of the fluid and hasten its demise.

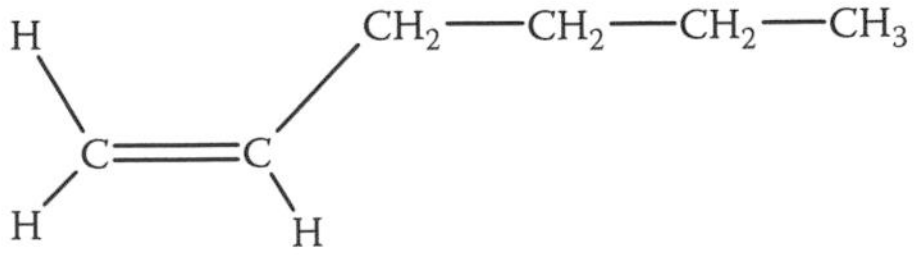

FIGURE 7.7 Structure of 1-hexene.

TABLE 7.4

Advantages and Disadvantages of Synthetic Oil

Advantages	Disadvantages
Better low- and high-temperature viscosity characteristics	High cost per quart
Better cold starts	Increased wear during break-in
Less evaporative loss	Possible compatibility with POM parts
Greater life	
Improved fuel economy	

Thermal considerations are paramount for ATFs. A great deal of friction is produced, which generates heat and high-temperature spikes at the torque converter; this will damage the fluid. ATF is designed to operate at temperatures around 95°C. At the torque converter, under extreme conditions, temperatures can reach 120°C. A transmission fluid should be designed to handle these temperature extremes. The transmission fluid will quickly break down above 120°C. Temperature spikes in the torque converter have been known to go above 120°C.

The torque converter is a type of fluid coupling device that hydraulically connects the engine to the transmission—analogous to a mechanical clutch. Used in conjunction with the torque converter is a stator, which essentially assists at low engine speeds, thus increasing acceleration. The vanes inside the converter alter the shape of the fluid path into the stator. The stator captures the kinetic energy of the transmission to enhance torque multiplication. This process will not only increase heat, but also increase shear of the transmission fluid. In addition to torque conversion, at every shift event, clutch packs generate heat, which must be carried away by the transmission fluid.

Transmission fluids are generally base oils from groups II or III with additive packages designed to improve oxidative stability, reduce foaming, reduce wear, and inhibit corrosion. Automatic transmission fluid should be replaced every 2–3 years or 24,000–36,000 miles of driving. Other hydraulic fluids include differential oils or gear oils used in manual transmissions. They utilize extreme pressure (EP) additives as well as antiwear and antifoaming additives.

7.13 SOME TESTING METHODS

Many tests are utilized to determine breakdown of these fluids, but we will not go into detail here. One of the most useful is to determine the total base number (TBN), which is a measure of the reserve alkalinity of an oil to neutralize acids. It is a wet method utilizing traditional laboratory techniques. The results are given as milligrams of KOH (potassium hydroxide) per gram of lubricant. The total acid number (TAN) is a measure of the lubricant's acidity. Other tests include an infrared analysis that can measure the increase in the carbonyl group at around $1,730\ cm^{-1}$ absorbance. This peak will grow as the oil oxidizes more. Another outstanding method is to utilize reverse phase high performance liquid chromatography (HPLC) if one has knowledge and the additive packages present. Another test is the basic sulfur, zinc, and phosphorus test.

7.14 TRANSMISSION FLUID TYPES

The many different types of transmission fluids can be somewhat confusing. Original equipment manufacture requirements should be met when replacing a transmission fluid. Using the wrong type of fluid may cause transmission problems and damage. Table 7.5 shows transmission fluid types.

TABLE 7.5
Transmission Fluid Types

Type F	Introduced by Ford in 1967; used by Toyota
Type CJ	Fluid for Ford C6 transmissions; incompatible with type F; compatible with Mercon and Mercon V
Type H	Ford specification compatible with Mercon and Mercon V
Mercon	Introduced in 1987; incompatible with type F; ceased production in 2007; serviced with Mercon V
Mercon V	Introduced in 1997 with Ford Ranger, Explorer, Aerostar, etc.
Dexron	Original GM for ATF
Dexron II	Improved GM formula for better viscosity and oxidation inhibitors; compatible with Dexron
Dexron IIE	ATF for electronic transmissions from GM
Dexron III	Replacement for Dexron IIE; improved oxidation and corrosion control for electronic transmissions
Dexron III (H)	Released in 2003; improved version of Dexron III
Dexron III/Saturn	Specialty ATF for Saturn products
Dexron VI	Released in 2006 for 6L80 6 speed RWD transmission; compatible with 2005 transmissions with Dexron III; not recommended for older transmissions
Chrysler 7176 (ATF+2)	Chrysler FWD transmissions
Chrysler 7176D (ATF+2)	Introduced in 1997; improved cold temperature flow and oxidation resistance
Chrysler 7176E (ATF+3)	Greater shear stability and improved base oil; incompatible with Dexron or Mercon
Chrysler ATF+4	Introduced in 1998; synthetic base introduced as a replacement for ATF+3; must be used in vehicles that were originally filled with ATF+4
BMW LT71141	BMW special formulation
Honda ZL	ATF for Honda automatics
Mitsubishi Diamond SP-II & SP-I11	Special formula for Mitsubishi
Nissan J-Matic	Special formula for Nissan
Toyota Type T, T-III, and T-IV	Toyota, ATF, and Lexus transmissions

7.15 ENGINE COOLANT

Engine coolants used in internal combustion engines can be considered to be cryoprotectant or colligative agents. Cryoprotectants are basically agents added to water to reduce the freeze point of a mixture. Colligative agents describe the behavior of coolant in warmer environments because they have beneficial behavior on both the freezing and boiling ends of the spectrum. It is these antiboil and antifreeze elements that the chemist seeks when he or she attempts to design an engine coolant.

Water, of course, was the first material used as an engine coolant. However, alcohols such as methanol and glycols were soon utilized as a standard for engine coolant. In 1926, ethylene glycol became available as "permanent antifreeze." Ethylene glycol's higher boiling points increase the operational range of the coolant.

7.16 METHANOL

Methanol has the structure shown in Table 7.6. It has been used as an antifreeze in the past, but is currently used in windshield wiper fluid as an additive. It is a light, volatile, flammable liquid at room temperature with a distinctive odor mildly different from that of ethanol.

7.17 ETHYLENE GLYCOL

Ethylene glycol is the major component used today as an antifreeze. It has a boiling point of 197.3°C. This higher boiling point is an advantage over compounds such as

TABLE 7.6
Various Coolant Properties

Compound	Boiling point (°C)	Mass (g/mol)	Density (g/cc)	Structure
Methanol	64.7	32.04	0.79	
Ethylene glycol	197.3	62.07	1.11	
Propylene glycol	188.2	79.06	1.04	
Diethylene glycol	244	106.12	1.12	

methanol and provides advantages for higher as well as lower ambient temperatures. Ethylene glycol is poisonous and must be disposed of through proper means. This compound forms calcium oxalate crystals in the kidneys and can cause acute renal failure. A bittering agent (denatonium benzoate) is added to coolant to discourage oral ingestion.

7.18 PROPYLENE GLYCOL

Propylene glycol is less toxic than ethylene glycol and is often labeled as a nontoxic antifreeze. This material is used in arenas other than the auto industry, such as water pipes and food processing. Propylene glycol will break down more readily and yield organic acids (formic, carbonic, and oxalic) upon exposure to excessive heat and air. The material will turn reddish in color upon breakdown.

7.19 NEW DEVELOPMENTS

Antifreeze contains corrosion-inhibiting compounds and dye for identification purposes. Dilution is 1:1 glycol to water mixture, which gives protection down to −40°C. Newer developments have seen the advent of organic acid technology (OAT) antifreezes such as Dex-cool. OAT antifreezes have extended service lives of up to 150,000 miles. The properties of OAT coolant are shown in Table 7.7. OATs can mix with non-OATs; however, the change interval will decrease to 2 years or 30,000 miles. OATs work by a mechanism different from that of traditional antifreeze mechanisms. As can be seen from Table 7.7, the additive package of the OAT contains

TABLE 7.7
OAT Coolant Properties

Physical	
Color	Orange
Odor	Sweet to faint
pH	8.0–8.6
Vapor pressure	<0.01 mm Hg at 20°C
Boiling point	108.9°C
Freezing point	−36.7°C
Flash point	127°C
Autoignition temperature	400°C
SpG	1.12 at 15.6°C
Viscosity	8 cSt at 40°C

Chemical	
Ethylene glycol	80–97% wt
Diethylene glycol	1–5% wt
2-Ethylhexanoate	1–5% wt

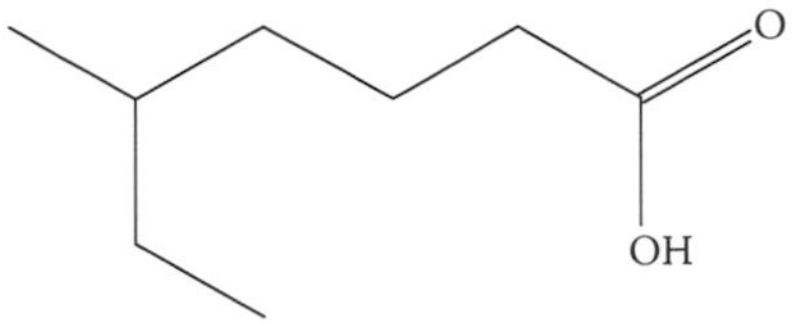

2–ethylhexanoic acid

FIGURE 7.8 Structure of 2-ethylhexanoic acid.

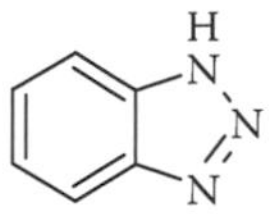

Tolyltriazole 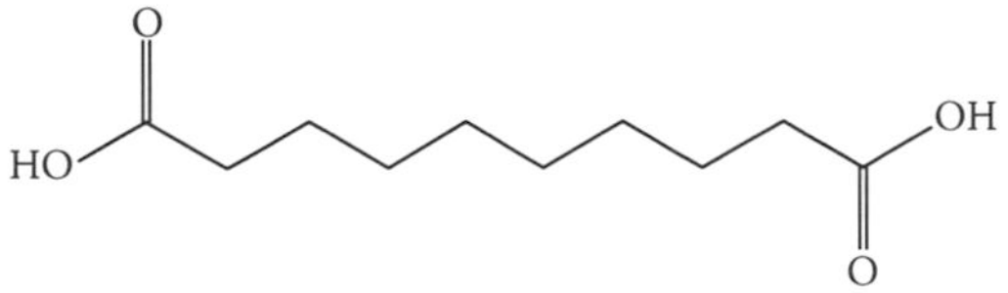   Sebacic acid

FIGURE 7.9 OAT additives.

1–5% 2-ethylhexanoate. Ethylhexanoate (Figure 7.8) is one of several additives that can be used with OATs.

Sebacic acid and tolyltriazole (shown in Figure 7.9) are also additives used in OATs. An organic acid additive package works by combining with metal constituents of a vehicle, thus providing a longer life. All new parts in the loop, such as water pumps, heat exchangers, coolant lines, and connectors, must be coated and a minimum concentration of fluid maintained. In addition to longer life, other benefits include

- lower alkalinity;
- no silicates (longer lasting water pump and engine seals);
- no nitrates, amines, borates, or nitrites; and
- biodegradability

Havoline obtained the original patent for OAT coolant. Newer OATs claim to be compatible with all types of coolants [5].

REFERENCES

1. Chenier P. J. 1992. *Survey of industrial chemistry,* 127, 130. New York: VCH.
2. Kister, H. Z. 1992. *Distillation design,* 1st ed. New York: McGraw–Hill.
3. Speight, J. G. 2006. *The chemistry and technology of petroleum,* 4th ed., 761. Boca Raton, FL: CRC Press.
4. Svara, J., N. Weferling, and T. Hofmann. 2006. Phosphorus compounds, organic. In *Ullmann's encyclopedia of industrial chemistry.* Wienheim, Germany: Wiley-VCH.
5. www.getahelmet.com/jeeps/maint/dexcool/

8 Seal and Gasket Design

8.1 INTRODUCTION

The importance of the chemist's role in automotive design is illustrated well by the necessity for interaction between the mechanical and chemical fields in sealing and gasket design. The chemist's knowledge of material and behavior under thermal and mechanical loads plays a key role in the design of these compounds. Maximum service use temperatures, compression set, flexural modulus, tear strength, dielectric properties, chemical resistance, ozone resistance, etc. are just some of the properties that must be considered by the material engineer or automotive chemist when designing a sealing system. Obviously, compression, tear, and service use temperatures have a higher weighted importance in seal and gasket applications. Seals and gaskets prevent fluid transfer between different regions of an operating system.

Key in choosing a material is the environment in which the seal will have to function. A chemist will typically have to ask questions such as:

What temperature extremes will the seal or gasket experience?
What continuous use temperature will be required?
How much relative movement will be between the two sealing surfaces?
Will the seal or gasket be exposed to chemicals, ozone, weather factors, etc.?
If exposure to chemicals occurs, what functional groups will interact with
     the seal?
What are the compression set and the performance expectations?

After answering these questions about the environment and performance requirements, the chemist can examine data from the various databases of material choices and make an informed decision. A successful seal design will ensure adequate compressive force while minimizing stresses acting upon the seal or gasket.

8.2 TEAR STRENGTH

One of the major considerations that must be taken into account is the tear strength of the material. Shear forces are often applied by the two surfaces being sealed (Figure 8.1). In this situation, the automobile chemist must take into account the forces being applied and the temperature at which the shear is occurring.

Figure 8.2 shows a diagram of tear versus temperature for some common elastomer compounds used in the automobile industry. The slope of the diagram should be considered if one is operating on a wider range of service temperatures. For instance, if an

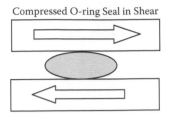

Compressed O-ring Seal in Shear

FIGURE 8.1   Diagram of compressed O-ring in shear.

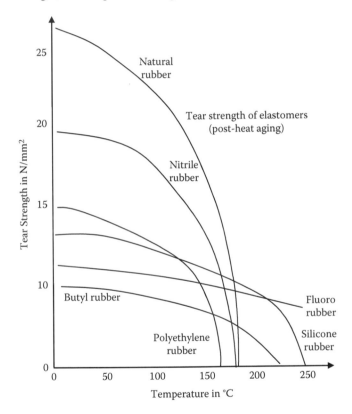

FIGURE 8.2   Tear strength of elastomers. (From Tomanek, A. *Silicones and Industry,* Wacker-Chemie GmBH, Munich, 1984. With permission.)

application approaches 150°C in continuous use and tear is a factor, then it would be preferable to shy away from natural rubber, nitrile, or PE rubber as a choice for the seal.

8.3   THERMAL SERVICEABILITY RANGE

The thermal serviceability or usage range is the temperature range in which a sealing material can be used and expected to function without degradation of properties. Table 8.1 shows the ranges of some compounds typically used in automotive sealing.

TABLE 8.1

Thermal Serviceability of Elastomers

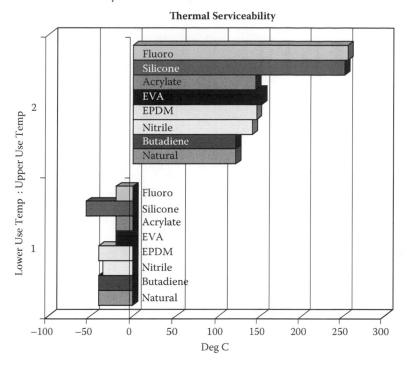

8.4 COMPRESSION SET

Another important characteristic the chemist must consider in seal and gasket design is the compression set, which is a measure of the ratio of elastic to viscous components of an elastomer to a given deformation [1]. Compression set is measured after a given load is applied for a certain time and temperature. It is measured as a percentage and can be considered to be the percentage of the original size not recovered. The testing of this property is usually conducted on cylindrical disks or O-rings. Along with the differential scanning calorimetry (DSC) techniques mentioned in Chapter 2, this technique is a good indicator of the amount of cure in a compound. If we look at Figure 8.3, we can see the compression set for various sealing materials.

The longer the polymer chain is, the better is the resistance to compression set because of the improved ability to store energy. In Figure 8.3, the sealing compounds are NBR (nitrile rubber), EPDM (ethylene–propylene diene rubber), FVMQ (fluoro-vinyl-methyl [fluorosilicone]), VMQ (vinyl-methyl silicone), FKM (fluroelastomers), and FFKM (perfluoroelastomers).

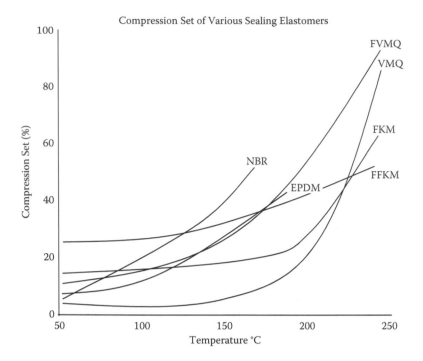

FIGURE 8.3 Compression set of various elastomers.

8.5 SILICONE RUBBERS

Silicone rubbers play a huge role in automotive applications due to their low temperature characteristics, thermal and chemical resistance, and general purpose use. From Table 8.1, we can see that the useful temperature range goes from –50 to 250°C. For automotive applications, this range covers all the operating system temperatures that a vehicle will see. Table 8.2 shows service life at continuous use for silicone rubbers. The time durations shown here are extensive for an elastomer and represent a great improvement over unsaturated carbon-based elastomers.

TABLE 8.2
Service Life of Silicone Rubbers
at Different Temperatures

Temperature (°C)	Time
150	2–4 years
200	1 year
250	100 days
300	14 days

The main types of silicone rubbers used in the automotive industry for sealing are HTV (high-temperature vulcanizing), RTV (room-temperature vulcanizing), liquid rubber, and RTV-2 (a two-component rubber used in cure-in-place applications) [2]. Figures 8.4 through 8.6 show these silicone rubbers. Of note in the structures is the Si–O–Si bond. The symmetry of the oxygen in this bond allows for rotation about the bond, giving the molecule great flexibility.

Table 8.3 shows some of the properties of the silicone rubbers. From the table, we can see that the RTV silicone rubbers typically have chain lengths of around 200 silicone units with molecular weights of around 20,000. The significance, however, is that the RTV-2 silicone rubbers are utilized much more in cure-in-place gasket applications and have seen increased use in recent applications. In cure-in-place applications, a bead of silicone is placed along the surface that is to be sealed. The part is then taken through a curing oven, usually set at around 250°F, for approximately 20 minutes. This process has a major advantage in that the gasket is adhered to the surface of the part, which eliminates rolling of a square-cut style gasket (see Section 8.12).

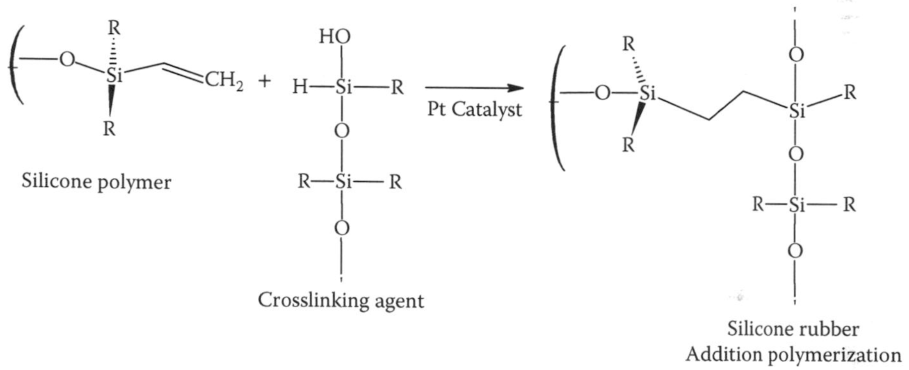

FIGURE 8.4 Silicone rubber from addition polymerization.

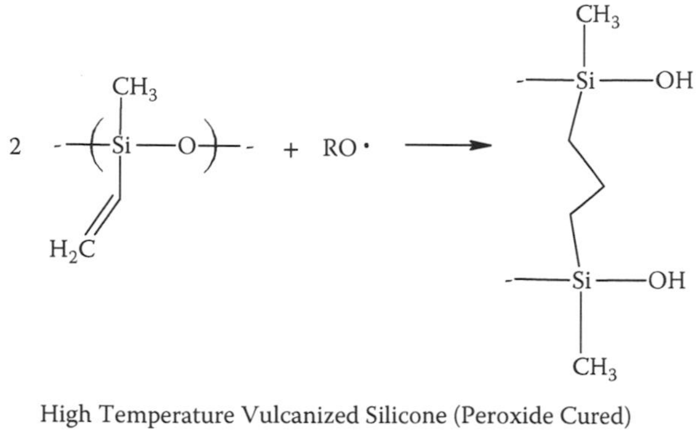

FIGURE 8.5 Silicone rubber from peroxide cure.

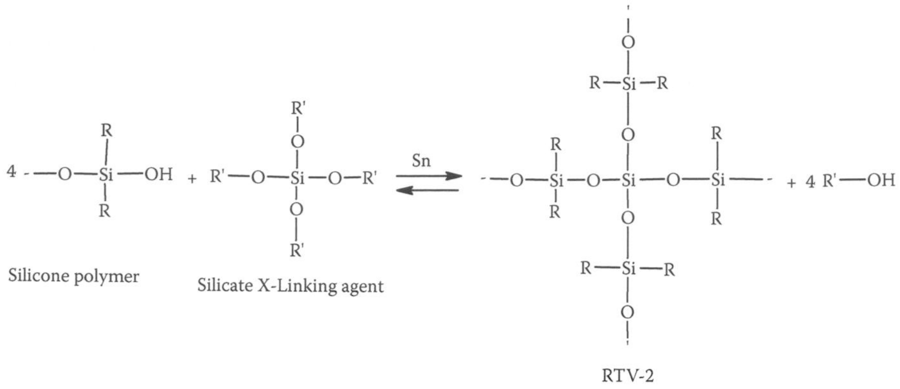

$$
4\ \text{---O---Si(R)(R)---OH} + \text{R'---O---Si(OR')(OR')---O---R'} \xrightarrow{\text{Sn}} \text{---O---Si---O---Si---O---Si--- } + 4\ \text{R'---OH}
$$

Silicone polymer　　　　Silicate X-Linking agent　　　　　　　　　　　　　RTV-2

FIGURE 8.6　Room temperature vulcanized (RTV) silicone rubber.

TABLE 8.3
Properties of Various Silicone Rubbers

Properties	Solid Rubber (HTV rubber)	Liquid Rubber	RTV Rubber
Polymer viscosity	20 million mPa s	5,000–100,000 mPa s	200–10,000 mPa s
Molecular weight	400,000–1 million	10,000–100,000	20,000
Chain length (silicone units)	10,000	1,000	200
Cross-linking mechanism	Peroxide at elevated temps.	Addition at elevated temps.	Condensation at room temp.

Source: Tomanek, A. *Silicones and Industry,* Wacker-Chemie GmBH, Munich, 1984. With permission.

TABLE 8.4
Usage and Temperature Range of Various Silicone Rubbers

Polymer Type	Symbol	Temperature Range (°C)	Notes
Dimethyl	MQ	−70 to 250	High comp. set; low thermal stability under load; high peroxide concentration during vulcanization
Vinylmethyl	VMQ	−70 to 250	Low comp. set; greater thermal resistance under permanent load; better vulcanization behavior
Phenylvinyl-methyl	PVMQ	−100 to 250	Good low-temp. resistance; better flame retardancy; lower oil resistance
Fluorovinyl-methyl(fluorosilicone)	FVMQ	−60 to 180	Great resistance to oils and aromatic solvents; low comp. set

Source: Tomanek, A. *Silicones and Industry,* Wacker-Chemie GmBH, Munich, 1984. With permission.

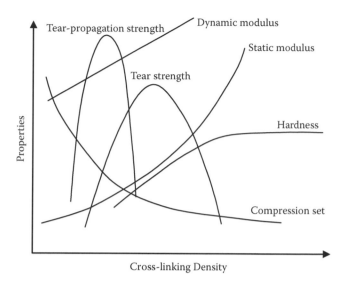

FIGURE 8.7 Cross-link density versus properties for silicone rubbers. (From Tomanek, A. *Silicones and Industry,* Wacker-Chemie GmBH, Munich, 1984. With permission.)

The type of "R" groups or polymer types along with their symbols and useful temperature ranges are listed in Table 8.4. Again we see the useful ranges can go well below the −40°C that is required for automotive applications. As the base polymer type, PVMQ (phenylvinyl-methyl) can go as low as −100°C.

Figure 8.7 shows the properties versus cross-link density for silicone rubbers. This chart shows the care that must be taken when choosing a material. For instance, tear strength is optimized at a certain cross-link density, but falls off rapidly when cross-linking increases. Hardness will plateau after a certain point along the x-axis, which may not give the tear strength required. Of course, if tear strength is the most important factor that the chemist is looking for, adjustments in filler content can be made to give the durometer required.

In addition to the flexibility of the oxygen–silicon bond, the effect of hydrogen bonding by fumed silica with the silano groups has been observed [2]. The interaction between the reinforcing filler (fumed silica) and the oxygen on the α,ω-dihydroxy polydimethyl siloxane increases viscosity and changes the T_g and crystallization temperatures [2]. Of course, siloxane polymers also contain reinforcing fillers such as chalk, quartz, mica kaolin, $Al(OH)_3$, and Fe_2O_3. These fillers are around 0.1 μm and they raise the viscosity of the non-cross-linked polymers as well as increase the hardness and modulus [2].

8.6 EPDM

Ethylene–propylene diene monomer (EPDM) rubbers are used in mass predominantly in isolation systems such as CRFM (condenser, radiator, fan module) or engine mounting. More varied are the sealing applications, which include transmission seals and o-rings, HVAC module seals and gaskets, radiator seals, weather stripping

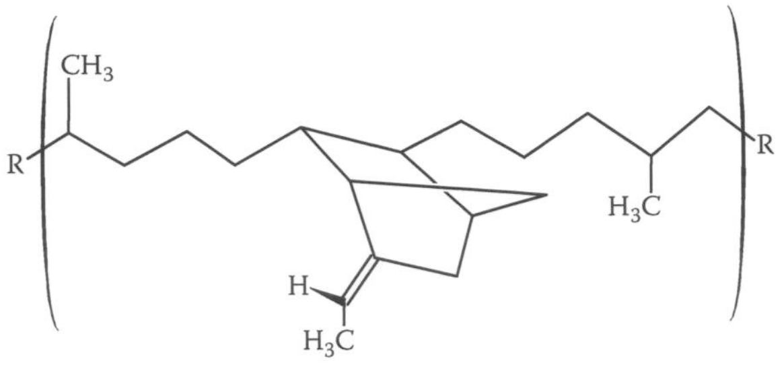

FIGURE 8.8   Structure of EPDM.

seals, and glass run channels, among other applications. In 2000, production figures for this synthetic polymer were over 870 metric tons [3]. The structure of EPDM is shown in Figure 8.8.

EPDMs are utilized in the automobile industry for their heat resistance, ozone resistance, and overall stability. Several key characteristics that are useful to the automotive chemist can be derived from the structure. Figure 8.8 shows the straight chain backbone of the EPDM polymer. EPDM is a terpolymer, or three-monomer polymer, consisting of propylene, ethylene, and another constituent—in this case, ENB (ethylidene norbornene).

Ethylene and propylene monomers combine to form the saturated and stable backbone [4]. This saturated backbone will provide excellent heat, oxidation, ozone, and weather aging because no reactive double bonds are in the backbone structure [4]. The third monomer (ENB) is added in a controlled manner and provides a site for cross-linking via the double bond. The M in EPDM refers to a saturated backbone. By virtue of the ENB, various amounts of vulcanization can be obtained to acquire the durometer, tear strength and tensile and other properties needed for the sealing or isolation needs of the automobile part.

Another of the monomers used in addition to ENB is dicyclopentadiene (DCPD), which is shown in Figure 8.9. As can be seen from the structure, there are two double bonds in DCPD. Each of the two dienes will have different tendencies for long chain branching, which will influence processing rates and cross-linking by sulfur or peroxide cures [3]. Table 8.5 shows some of the characteristics of the two dienes [5]. In addition, Table 8.6 shows general features of ethylene–propylene elastomers as related to the ter-

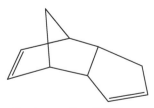

Dicyclopentadiene DCPD

FIGURE 8.9   Structure of dicyclopentadiene.

TABLE 8.5
Properties of EPDM Dienes

Termonomer	Cure and Property Features	Long Chain Branching
ENB	Fastest and highest state of cure good tensile; good compression set resistance	Low to moderate
DCPD	Slow sulfur cure; good compression set resistance	High

TABLE 8.6
Features of Ethylene–Propylene Elastomers

Characteristics	High	Low
Ethylene Content	Good green strength	Fast mixing
	Flow at high extrusion temps	Low-temp. flexibility
	High tensile strength, modulus	Low hardness and modulus
	High loading (reduced cost)	Calendering and milling
Diene content	Cure degree and fast rate	Scorch resistance
	Acceleration versatility	High-heat stability
	Good compression set	Low hardness and modulus
	High modulus, low set	
Molecular weight	Good tensile, tear modulus, set	Fast mixing
	High loading and oil extension	High extrusion rates
	Good green strength	Good calendering
	Collapse resistance	Low viscosity, low scorch
MWD	Overall good processing	Low die swell
	Extrusion feed and smoothness	Fast extrusion rate
	Collapse resistance	High cure
	Good calendering and milling	Good physicals

monomer content [4]. Using a table like this, the material engineer or chemist who needs good compression set resistance may want to choose a high diene content in the EPDM.

Ethylene–propylene elastomers are one of the main choices of automotive chemists and material engineers for sealing applications, as well as isolation systems, weather stripping, electrical components, etc. Heat aging resistance up to 130°C can be expected with certain sulfur cures and up to 160°C with peroxide-cured compounds [3]. They also respond well to high filler and plasticizer loads. Typical properties are shown in Table 8.7 [4].

TABLE 8.7
Properties of Ethylene–Propylene Elastomers

Polymer Properties	
Mooney viscosity, ML 1–4 at 125°C	5–200+
Ethylene content (wt. %)	45–80
Diene content (wt. %)	0–15
Specific gravity (g/mL)	0.855–0.88

Vulcanizate Properties	
Hardness (Shore durometer)	30A–95A
Tensile strength (MPa)	7–21
Elongation (%)	100–600
Compression set B (%)	20–60
Useful temperature range (°C)	−50 to +160
Tear resistance	Fair to good
Abrasion resistance	Good to excellent
Resilience	Fair to good
Electrical properties	Excellent

8.7 NATURAL RUBBERS

Natural rubber is utilized mainly as an isolation material in CRFMs in HVAC components. Natural rubber is tapped from the *Hevea brasilienis* tree in the form of latex or a colloidal suspension. These trees produce more rubber when they are wounded, thereby providing a renewable source of rubber. Figure 8.10 shows the two base components present in natural and synthetic rubbers. The repeating units consist of isoprene and some diene monomer such as butadiene shown in the figure and natural impurities. Synthetic rubber is typically made with isoprene and butadiene.

Due to the structure of natural rubber, with its conjugated system of double bonds and the ability to reduce the unwanted dynamic motions associated with a mounting system, isolation is its most important use. However, natural rubber use in seals and gaskets is a natural fit. The automotive chemist chooses natural rubber for several reasons. The elastic behavior of rubber can be attributed to electrostatic strain

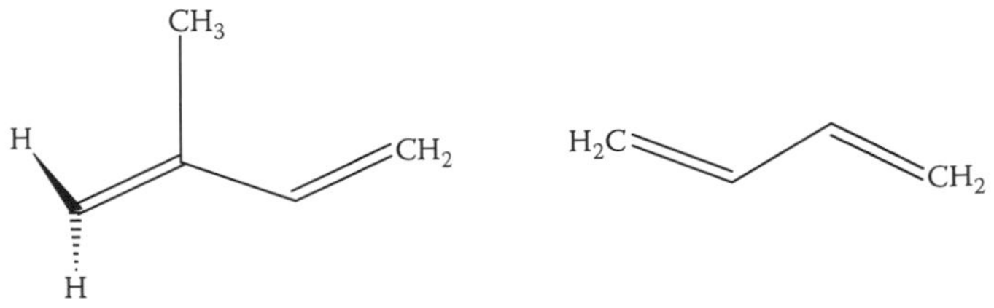

2-Methyl-buta-1,3-diene
(Isoprene)

1,3,-Butadiene

FIGURE 8.10 Structure of isoprene and 1,3-butadiene.

within the molecule, with the bonds being stretched beyond their minimum energy capacity, but a thermal component is also involved. Again, rotation of a molecule plays a role because, between cross-linked sites, different geometries can occur as the molecule rotates. The molecular interactions at room temperature store kinetic energy as oscillations occur. When the structure is strained, this energy is released in the form of heat and a decrease in entropy occurs. We can see this relation in the following equation:

$$S = k_B ln\Omega$$

(8.1)

Here, S is entropy, k_B is the Boltzmann constant (1.38066×10^{-23}), and Ω is the number of microstates that the molecule can assume per macrostate. If we consider entropy as the inability of a system's energy to do work, we can see that entropy will decrease as the chain is stretched [6]. Likewise, as the chain is relaxed, entropy will increase in an endothermic process.

Natural rubber shows a strong dependence on temperature; it has behavioral characteristics in three regions: glassy region, transition region, and rubber region [7]. At lower temperature, the rubber will be crystalline and, as the rubber's temperature increases, it enters the transition phase where it becomes leather-like [7]. Eventually, the material will enter the rubber phase with the sheer modulus decreasing in each phase [7].

As natural rubber is vulcanized, the disulfide bonds shown in Figure 8.11 shorten the chains of the rubber and increase the rate at which the chain will contract. The greater the number of disulfide bonds, the greater is the hardness of the natural rubber. Hardness affects the seal's ability to compress as well as its performance in thermal cycling events.

FIGURE 8.11 Diagram of disulfide bond.

As with all rubbers, natural rubber (NR) is greatly affected by the type of filler utilized, amount of filler, and shape of the filler. The filler content contributes to the Payne and Mullins effects. The Payne effect is exhibited in seals and gaskets with carbon black filler. A. R. Payne studied this effect, which is seen during cyclic loading conditions upon a rubber material [8]. In this effect, as the strain gets larger, the storage modulus decreases. The effect is seen at approximately 0.1% strain amplitude [8]. That is, at over 0.1% strain amplitude, the ability of a rubber to store energy decreases rapidly. At 20%, a lower limit comes into play [8]. The Payne effect is not observed in samples without filler. Likewise, the Mullins effect (named for Leonard Mullins) is the stress–strain phenomenon that can be considered softening of the stress–strain behavior when a load is beyond that which has been previously reached.

In 2005, nearly 8.6 million tons of natural rubber were produced. Of this amount, 94% was from Thailand, Malaysia, and Indonesia. The total world consumption of rubber is around 18 million tons per year with about 20% SBR (styrene butadiene rubber), 14% latex SB, 12% polybutadiene, 5% EPDM, 2% polychloroprene, 2% nitrile, and 7% other synthetics.

Demand for elastomers—for synthetic rubber (SR) as well as NR—is well secured and is continuously increasing at a rate of 3–4% per year [9]. Because rubber is a petroleum-derived product manufactured by a polymerization process in chemical plants, the management of supply against demand is relatively straightforward. To a certain extent, the prices of its basic ingredients—namely, the monomers—are more or less influenced by the price of petroleum [9]. It is important for the imperative of cost that we consider this fact when deciding between natural and synthetic substances.

8.8 NITRILE RUBBERS

Nitrile rubber is another synthetic rubber composed of acrylonitrile and butadiene (Figure 8.12). Most automotive o-ring applications are made of this copolymer. Nitrile rubber is also known as NBR or buna-N. One of the chief advantages of NBR is its resistance to fuel as well as oil. In the acrylonitrile portion of the polymer, the arrangement of the polar nitrogen group is shown in Figure 8.13. In addition to the ability of the rubber to arrange itself in a *cis* or *trans* configuration, this group allows for repulsion of fuel and oil to take place. The higher the amount of acrylonitrile in the polymer, the greater the oil resistance of the polymer is [10,11]. The flexibility of the chain will be greatly affected, however, as the amount of ACN rises. As one can imagine, not only is NBR used for seals and gaskets but also for fuel and oil lines. It should be mentioned that the conjugation in nitrile rubber (as well as natural rubber) allows for attack by ozone, ketones, esters, aldehydes, and aromatics. Hydrogenating an NBR will reduce the chemical reactivity of the polymer backbone [11, Table 46 therein].

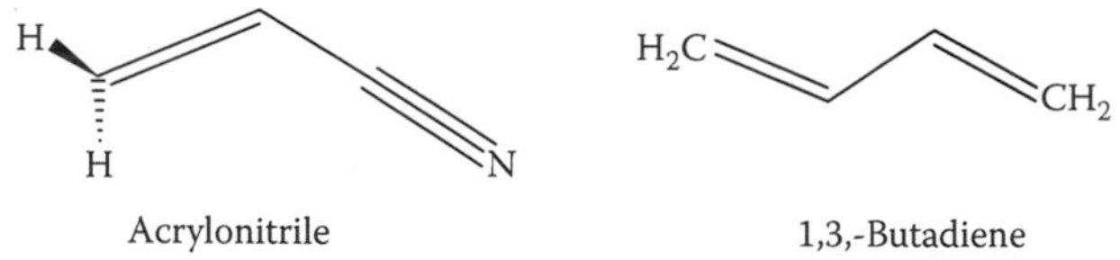

Acrylonitrile   1,3,-Butadiene

FIGURE 8.12   Structure of acrylonitrile and 1,3-butadiene.

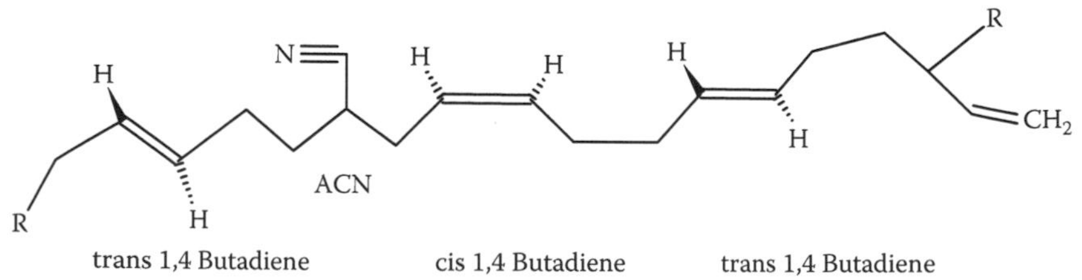

trans 1,4 Butadiene   cis 1,4 Butadiene   trans 1,4 Butadiene

FIGURE 8.13   Diagram of acrylonitrile showing *cis*- and *trans*-butadiene.

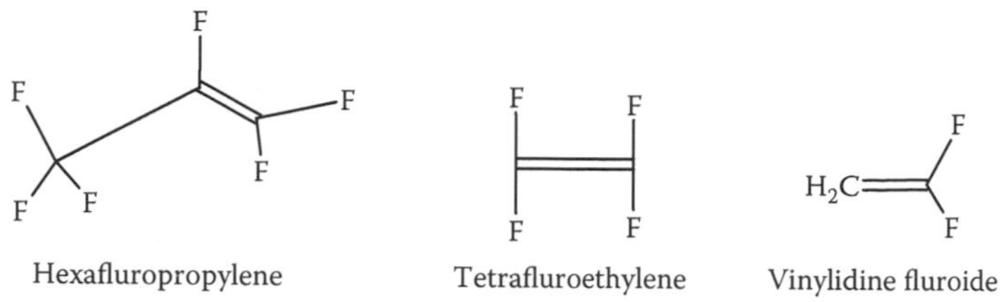

Hexafluropropylene Tetrafluroethylene Vinylidine fluroide

FIGURE 8.14   Structures of fluoroelastomers.

8.9   FLUOROPOLYMER ELASTOMERS

Fluoropolymer elastomers (or Viton, a registered trademark of the DuPont Performance Elastomer LLC) consist of hexafluoropropylene (HFP), vinylidene fluoride (VDF), and tetrafluoroethylene. The 3M Corporation uses the trade name Fluorel. The structures of each of these monomers are shown in Figure 8.14. The fluorine content of these terpolymers is typically around 70% for Viton. Four basic types of this material are shown in Table 8.8; however, Viton Extreme comprises more types.

Viton has great resistance to oils, solvents, and other petroleum products. High-temperature performance is outstanding. The semiconductor industry has a higher use rate of Viton due to its electrochemical properties [12]. Viton can retain half of its original properties after 16 hours at 600°F. It is resistant to oxidation but can be attacked by highly polar groups such as ketones, hydrazine, etc.

Several factors contribute to the high heat stability of these compounds. There are extremely strong bonds between the carbon atoms in the polymer backbone and the attached fluorine atoms [13]. These factors help the polymer resist chain scission. In addition, the high fluorine to hydrogen ratio and saturation of the backbone increase the strength and stability of that polymer backbone [13]. Table 8.9 shows some bond dissociation energies that must be exceeded to rupture the bond.

Looking at Table 8.8 we can see that Viton A is a general purpose type of material and is the most widely used. It is a copolymer of vinylidene fluoride (VF2)

TABLE 8.8
Properties of Fluoropolymer Elastomers

Type	Makeup	Uses
A	Vinylidene fluoride, hexafluoropropylene	General sealing; automotive, lubricants
B	Vinylidene fluoride, hexafluoropropylene, tetrafluroethylene	Chemical process plants; power utility seals and gaskets
F	Vinylidene fluoride, hexafluoropropylene, tetrafluroethylene	Oxygenated automotive fuels; concentrated aqueous inorganic acids; water; steam
Viton Extreme	Tetrafluroethylene, propylene and ethylene, polymethyl vinylidine	Automotive; oil exploration; special sealing; ultraharsh environments

TABLE 8.9
Bond Dissociation Energies

Bond	Dissociation Energy
–C–C–	83
–C–H	91
–CH–H	95
–CH$_2$–H	98
–CH$_2$–O–	106
–CH$_2$–F–	93
–C–N–	82
–Si–O–	106
–CH–Cl	78

and hexafluoropropylene (HFP) [13]. Viton B is a terpolymer of tetrafluoroethylene (TFE), VF2, and HFP. These compounds have about 68% fluorine levels and have improved fluid resistance as compared to type A [13].

Viton F is a terpolymer combining TFE, VF2, and HFP as well, with the fluorine level being raised to approximately 69%. Type F does not have good low-temperature capability. The Viton Extreme polymers consist of such compounds as Viton GF, GLT, and GFLT. Viton GF is composed of TFE, VF2, HFP, and a small amount of cure site monomer, which allows peroxide curing of the compound [13]. This type will have increased resistance to water-based compounds such as automotive coolants. The GLT type has improved low-temperature characteristics—thus the LT. The T_g of these materials is approximately 8–12°C lower than for type A [13]. Finally, the GFLT type combines VF2, perfluoromethylvinyl ether (PMVE), TFE, and a cure site monomer [13]. The idea here is to provide greater chemical resistance, high heat resistance, lower swell, and low-temperature capabilities.

8.10 ETHYLENE ACRYLIC SEALS

Ethylene acrylic seals are known as Vamac (another registered trademark of the DuPont Corporation). DuPont originally developed ethylene acrylic elastomers to improve the heat resistance of acrylonitrile and neoprene. These materials are terpolymers of ethylene, methyl acrylate, and a cure site monomer. They are curable with peroxides. The amount of methyl acrylate and ethylene varies. In Figure 8.15, we can see two monomers: one with maleic anhydride (MAH) and one with glycidyl methacrylate (GMA). These are resistant to oxygen and ozone, and they have superior heat resistance. The backbones of these polymers are saturated as well. There is a slight cost benefit when choosing one of these compounds over a fluorinated one. After a postcure, compression set is excellent with values that can go below 20%. Of all of the seal and gasket types mentioned here, it is up to the individual chemist or

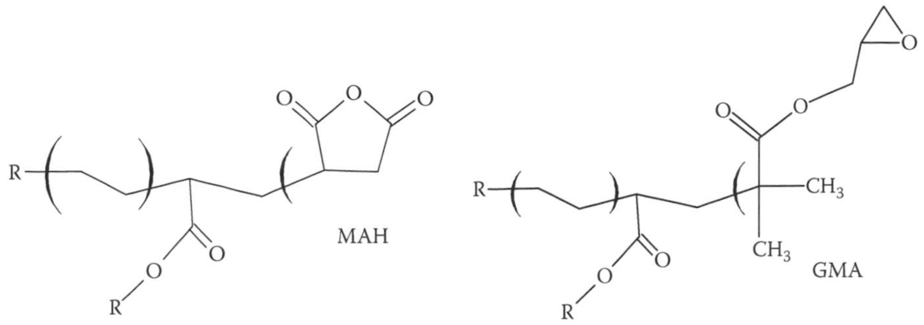

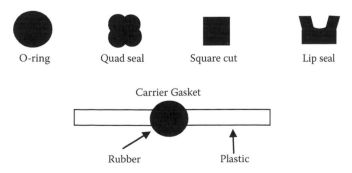

FIGURE 8.15   Ethylene acrylic monomers.

material engineer to evaluate each seal material type and choose the seal that best fits what he or she is trying to do.

8.11   POLYETHERKETONE (PEEK), POLYETHERIMIDE (PEI), AND TEFLON (PTFE)

These three compounds are considered engineering polymers and as such were discussed in more detail in Chapter 7. Their plastic, nonelastic behavior does not limit their ability as a seal. Dynamic sealing (reciprocating and rotating) is the main use for these materials.

8.12   SEAL TYPES

Numerous seal designs are available to the automotive chemist. Choices are made depending on the application and sealing requirements. Figure 8.16 shows a few of the types of seal designs used.

In seal and O-ring design, three considerations are typically examined: thermal mechanical, viscoelastic, and incompressibility [14]. Thermal mechanical is basically the material changes that we have been discussing under load. Incompressibility is exhibited when a material has an isochoric (zero volume) change under pressure. If

O-ring   Quad seal   Square cut   Lip seal

Carrier Gasket

Rubber   Plastic

FIGURE 8.16   Automotive seal types.

the Poisson ratio is slightly less than 0.5, then the material is nearly incompressible [14]. Viscoelasticity is rate-dependent behavior with properties that change with temperature and time. The features of a viscoelastic material are as follows [14]:

- creep under constant stress;
- relaxation under constant strain;
- hysteresis under loading and unloading cycles;
- internal friction rearrangement of molecular structures under load;
  - strain-induced crystallization formation and melting of crystallized regions;.
  - stress softening; and
  - breakdown of the reinforcing filler and polymer bonds.

FIGURE 8.17 Automotive seal in groove.

O-rings are basically a round loop of material that can be used as a mechanical seal or gasket. In automotive design, typically a 30% compression is maintained and the seal sits in some sort of groove, as in Figure 8.17. O-rings are used to prevent the transfer of fluid between or among two or more components. The compressive force action on the seal is relied upon to prevent the transfer of fluids. The seal must be able to handle the stress acting upon it [15]. For static seals, we typically think of flange seals, radial or piston seals, and crush seals. Dynamic seals are discussed more thoroughly in Chapter 7. In these static types, there is no relative motion between moving parts.

FIGURE 8.18 Automotive piston seal in groove.

Figure 8.18 shows a flange type of static seal. The figure shows a piston seal similar to the PEEK seals discussed in transmission applications in Chapter 7, but without the rotating member. Static crush seals are exactly that. They often employ a lip seal or quad seal type of design, which is good for sealing against the surface in two points.

8.13 FAILURE AND DEGRADATION IN SEAL DESIGN

Seal failures can be extremely detrimental to a vehicle system and customers' enthusiasm and willingness to purchase a product. A customer's time and money are invested in his or her purchase, and the failure of a minor component should be avoided by a good design. Good material selection by understanding the physical requirements and good design by understanding the seal interfaces can prevent failures. From the end-user's point of view, a seal can fail in three general ways: leaking, contamination, and change in appearance [16]. All polymeric seals will degrade within the environment that they are exposed to within their lifetime. A seal's losses in mechanical strength, aesthetic appearance, compression set, etc. are adversely affected by certain chemical reactions, some of which are discussed here. Various

Initiation:

ROOH $\longrightarrow$ RO* + HO*

2ROOH $\longrightarrow$ RO* + ROO* + H$_2$O

Propagation:

ROO* + RH $\longrightarrow$ ROOH + R*

R* + O$_2$ $\longrightarrow$ ROO*

FIGURE 8.19 Shelton kinetic scheme for rubber.

stabilizers are added for the prevention of this premature degradation. Degradation is the irreversible change that will eventually lead to a failure. Changes in mechanical properties result from reactions that change the size of molecules or change the cross-link density [17].

8.14 THERMAL DEGRADATION

If there are no external reactants and thermal degradation occurs, then the degradation can be considered pyrolysis. There are three mechanisms by which it occurs in this case: random scission, which is predominant in polyolefins; depolymerization, which occurs at chain ends; and elimination of low molecular weight fragments. Chain scission is a random occurrence in which larger constituents are broken into smaller molecules along the backbone of the chain. For seals and gaskets, depolymerization is not of great concern. With this mechanism, the end molecules will split off sequentially; however, it is most prevalent in polyacetals and polymethyl methacrylate [17]. Likewise, PVA and PVC will pyrolyze by elimination of low molecular weight fragments such as PVC losing hydrogen chloride. Addition of some copolymers such as polystyrene with polyisoprene will reduce the rate of chain scission. Polyisoprene will generate small radicals as it undergoes chain scission. As diffusion of the radicals occurs in the polystyrene phase, hydrogen abstraction will take place and stabilization will occur [17].

8.15 THERMAL OXIDATION

Thermal oxidation occurs when heat and oxygen are exposed to the seal. The term *auto-oxidation* is used for temperatures between ambient and 200°C. Shelton proposed the kinetic scheme for rubber shown in Figure 8.19. Autotermination by coupling or disproportionation of radicals will occur [18]. Hydrogen peroxide is assumed to be a source of radicals that will initiate oxidation in polymers [18].

REFERENCES

1. www.pspglobal.com/index.html
2. Tomanek, A. 1991. *Silicones and industry,* 42. Munich: Wacker-Chemie GmbH.

3. Riedel, J. A., and R. Vander Laan. 1990. Ethylene propylene rubbers. In *The Vanderbilt rubber handbook,* 13th ed., 123–148. Norwalk, CT: R. T. Vanderbilt Co., Inc.

4. Karpeles, R., and A. V. Grossi. 2001. EPDM rubber technology. In *Handbook of elastomers,* 2nd ed., ed. A. K. Bhowmick and H. L. Stephens, 845–876. New York: Marcel Dekker, Inc.

5. Ver State, G. 1986. Ethylene propylene elastomers. In *Encyclopedia of polymer science and engineering,* vol. 6, 522–564. New York: John Wiley & Sons.

6. Daintith, J. 2005. *Oxford dictionary of physics,* 5th ed. Oxford, England: Oxford University Press.

7. Mattias, S. 2002. On dynamic properties of rubber isolators. Doctoral thesis. Royal Institute of Technology, Department of Vehicle Engineering, Stockholm, p. 8.

8. Payne, A. R. 1962. The dynamic properties of carbon black-loaded natural rubber vulcanizates. Part I. *Journal of Applied Polymer Science* 6 (19): 53–57.

9. http://www.fao.org/DOCREP/006/Y4344E/y4344e0d.htm

10. Mackey, D., and A. H. Jorgensen. 1999. Elastomers, synthetic (nitrile rubber). In *Kirk-Othmer encyclopedia of chemical technology,* 4th ed., 687–688. New York: McGraw–Hill.

11. Hoffmann, W. 1964. *Nitrile rubber, rubber chemistry and technology, a rubber review for 1963,* 154-160. San Francisco, CA: John Wiley & Sons.

12. http://www.marcorubber.com/viton.htm

14. http://www.pspglobal.com/static.html

15. http://www.pspglobal.com/viton2.html

16. http://www.pspglobal.com/seal-failure.html

17. Hawkins, W. L. 1984. Polymer degradation and stabilization, 5, 12, 15. Berlin: Springer–Verlag.

18. Shelton, J. R. 1972. Stabilization against thermal oxidation. In *Polymer stabilization,* ed. W. L. Hawkins, 29. New York: Wiley-Interscience.

9 HVAC System Overview and Refrigerant Design

9.1 INTRODUCTION

In the automobile industry, chemists and chemical engineers are often employed in heating, ventilation, and cooling (HVAC) departments due to their unique knowledge of such subjects as heat transfer, chemical behavior of refrigerants, chemical interactions between oils and other compounds, and the materials used in an HVAC module. In this chapter I would like to describe briefly an HVAC system so that we can understand the importance of the chemical knowledge around it. However, this is not intended as a detailed account on the subject. We will also discuss refrigerants and the importance of the chemistry of their design.

9.2 OZONE DEPLETION

In 1995 the Nobel Prize for chemistry was awarded to F. Sherwood Rowland and Mario Molina, physical chemists from the University of California, Irvine. In their published ozone depletion hypothesis [1], they proposed that chlorine atoms could form high in the stratosphere. As an offshoot of this work as well as some other work, the Montreal Protocol on Substances That Deplete the Ozone Layer was signed in 1987. The treaty took effect on January 1, 1989, and has since undergone several revisions. The treaty addresses ozone-depleting compounds that contain chlorine or bromine. Fluorine is not included in the treaty because it has not been shown to harm the ozone layer.

Ozone absorbs ultraviolet radiation from the sun while at the same time heating the gases in the stratosphere. Ozone depletion potential (ODP) is derived from its ability to deplete the ozone layer using trifluoromethane as a standard. Ozone absorbs short wavelength ultraviolet radiation from the sun; in the stratosphere, it is measured in Dobson units (DUs). One DU is equal to the amount of ozone that would be 10 μm thick under standard temperature and pressure. The average DU has been about 300 for the last few decades; however, this number has decreased in recent times.

Figure 9.1 shows ozone depletion over time as well as the corresponding UV increase. The UV index was taken at noon in Lauder, New Zealand, for consistency. The UV index is a measure of the relative strength of the sun; the intention is to give people an idea of how to protect themselves from the sun's radiation. A value of 10 represents a clear mid-day sun, and 0 represents no UV radiation at all. In some areas of Australia, values have been as high as 17 [2]. Typically, radiation is measured in

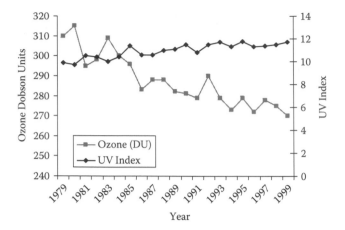

FIGURE 9.1   Ozone measurements in Dobson units per year.

watts per square meter (W/m^2). Because the wavelengths of concern are 295–325 nm (shorter wavelengths that damage the skin), the power spectrum is weighted according to the McKinlay–Diffey erythema action spectrum [2]. This spectrum roughly assigns a value from 1 to 10 for convenience.

9.3   MONTREAL PROTOCOL TREATY

Within the Montreal Protocol treaty, chlorofluorocarbons (CFCs) were broken up into several groups. Of primary importance was group 1 of annex A of the treaty, shown in Table 9.1. These CFCs are deemed to be more detrimental to the ozone layer on the basis of their ODP values. The ozone depletion potential is the relative amount of damage that a particular chemical causes to the ozone layer relative to R-11 (trichlorofluoromethane), which has an ozone depletion potential of 1.0.

The Montreal Protocol's purpose was that the signatory states [3]

> Recognizing that world-wide emissions of certain substances can significantly deplete and otherwise modify the ozone layer in a manner that is likely to result in adverse effects on human health and the environment,...[we are] determined to protect the ozone layer by taking precautionary measures to control equitably total global emissions of substances that deplete it, with the ultimate objective of their elimination on the basis of developments in scientific knowledge....Acknowledging that special provisions are required to meet the needs of developing countries...have agreed as follows...

According to the treaty, the CFC levels of consumption and production in group 1, annex A, from a period of 1991–1992 were controlled so as not to exceed 150% of calculated levels from 1986. From 1994, their calculated levels of consumption and production were not to exceed 25% of the 1986 value, and from 1996 the calculated level was not to exceed zero [3].

TABLE 9.1

Some Refrigerants and Their Designations

Chemical Structure	Refrigerant Designation
$CFCl_3$	CFC-11
CF_2Cl_2	CFC-12
C_2F_3Cl3	CFC-113
C_2F4Cl_2	CFC-114
C_2F_5Cl	CFC-115

Other CFCs, such as Halon 1211 and 1301 and chemicals such as CCl_4, will be phased to zero by 2010 [3]. Less active HCFCs will be completely phased out by 2030, with some exceptions in special cases. Since the inception of the treaty, there has been a steady decline in chlorofluorocarbons [4]. In 1985, Farman, Gardiner, and Shanklin published a study in *Nature* that showed a hole in the ozone layer [4a]. With evidence such as this, the auto industry is faced with being a better environmental steward. The major auto manufacturers have begun to utilize their chemists' talents not only in the design of lighter, stronger materials for HVAC units but also in the design of refrigerants.

9.4 REFRIGERANT DESIGN

Crucial to an HVAC is its refrigerant. The very natures of the various refrigerants used today drive the need for input from the chemist. Chief among them is the understanding of how a refrigeration system works. For automotive applications, the requirements differ from commercial and home requirements of an HVAC system. In automobiles, ambient conditions vary greatly, from the extreme cold of an Alaskan climate to the heat of Death Valley in the southwestern United States. In addition, the size of the system, insulation, solar load, etc., all come into play when designing a system. Table 9.2 shows some typical refrigerants and relevant information.

A brief discussion of key characteristics in refrigerant design is in order. The refrigerant designation was created by DuPont in 1956 [5]. The system basically designates the number of fluorine atoms with the first digit; the second digit represents the number of hydrogen atoms (+1), the third the number of carbon atoms (−1) or zero if there are no carbons, and the fourth is the number of unsaturated carbon–carbon bonds. The chemical name is the basic IUPAC compound name.

9.5 GLOBAL WARMING POTENTIAL (GWP)

Environmental concerns are extremely important when designing a refrigerant system and need to be taken into consideration. Several factors come into play and need to be mentioned. These are global warming potential (GWP), total equivalent warming impact (TEWI), and ODP.

TABLE 9.2
Typical Refrigerants and Their Properties

Designation	Chemical Name	GWP	ODP	BP (F)	Structure
R12	Dichlorodifluoromethane	8500	0.82	−21.6	
R134a	1,1,1,2-Tetrafluoroethane	1300	0	−15	
R152a	1,1-Difloroethane	120	0	−11.2	
R744	Carbon dioxide	1	0	−108	
R290	Propane	20	0	−43.8	
R600a	Isobutane	20	0	10.8	
RC270	Cyclopropane	20	0	−28.3	

GWP is the increase in average temperature of the Earth's surface and oceans seen over the past 100 years. It is estimated that, during that period, the temperature increase was approximately 0.74°. The Intergovernmental Panel on Climate Change (IPCC) has designated the likely cause as greenhouse gases [6]. GWP is measured using carbon dioxide as a standard gas and its value is derived from Equation 9.1 [6]:

$$GWP(x) = \frac{\displaystyle\int_0^{TH} a(x) * \left[x(t)\right] dt}{\displaystyle\int^{TH} a(r) * \left[r(t)\right] dt}$$

(9.1)

The IPCC defines global warming potential as follows [6]:

The GWP has been defined as the ratio of the time-integrated radiative forcing from the instantaneous release of 1 kg of a trace substance relative to that of 1 kg of a reference gas (IPCC, 1990): where TH is the time horizon over which the calculation is considered, a_x is the radiative efficiency due to a unit increase in atmospheric abundance of the substance in question (i.e., $Wm^{-2} kg^{-1}$), [x(t)] is the time-dependent decay in abundance of the instantaneous release of the substance, and the corresponding quantities for the reference gas are in the denominator. The GWP of any substance therefore expresses the integrated forcing of a pulse (of given small mass) of that substance relative to the integrated forcing of a pulse (of the same mass) of the reference gas over some time horizon. This value has supplanted Ozone Depletion Potential as far as concern over global effects on the environment.

9.6   TOTAL EQUIVALENT WARMING IMPACT (TEWI)

Total equivalent warming impact (TEWI) is the sum of direct as well as indirect contributions of global warming to the environment by greenhouse gases. For an automobile system, the refrigerant leaks directly into the system and energy is required to power the AC system. There is little opportunity to reduce the indirect emissions because the need to power the system from the automobile engine will always be present. An opportunity exists in the refrigerant choice—that is, replacing an R134a or similar system with a different refrigerant such as R152a, hydrocarbon, R744, etc. We can consider TEWI as having a "controllable" portion, which can be described as [7]

$$TEWI_{(controllable)} = (D_{R134a} - D_{alternative}) + (I_{R134} - I_{alternative})$$ (9.2)

Here, D is direct contribution and I is the indirect contribution to global warming. The direct portion of TEWI can be eliminated by replacement of R134a. Table 9.3 shows global warming impact in CO_2 per year for some refrigerants and the percentage of reduction by switching refrigerants.

9.7   OZONE DEPLETION POTENTIAL (ODP)

As can be seen from Table 9.2, ozone depletion values are zero for all compounds with the exception of R12. As stated earlier, the ozone depletion potential is the relative amount of damage that the particular chemical causes to the ozone relative to R-11 (Figure 9.2). Figure 9.3 shows the global warming impact of various refrigerant

TABLE 9.3
Global Warming Impact[a]

	R134a		R152a		R290		Sec. loop w/R152a		Sec. loop w/R290	
	Direct	Indirect	Direct	Indirect	Direct	Indirect	Direct	Indirect	Direct	Indirect
100°F × 40% R.H.	125	262	20	240	2	255	20	316	2	313
(TEWI)	125	127	130	51	72					
80°F × 50% R.H.	125	135	20	128	2	31	20	150	2	151
(TEWI)	125	112	126	90	106					
Av. (TEWI)	125	119	128	70	89					
Percent improvement of "controllable" TEWI	0	95	100	56	71					

Source: Goodbane, M. *SAE* Paper 1999-01-084, 1999. With permission.
[a] Kilograms of CO_2 per year.

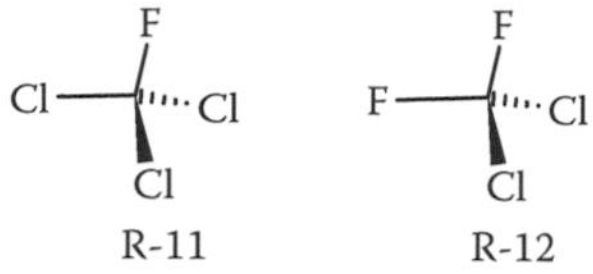

FIGURE 9.2 Structure of R-11 and R-12 refrigerants.

TABLE 9.4
Some Properties of Halogens

Halogen	Mass	Electronegativity	Ionization Potential (kJ/mol)	Electron Affinity (kJ/mol)
F	18.99	3.98	1681	328
Cl	35.45	3.16	1251	348
Br	79.9	2.96	1140	324

systems. A molecule of R11 contains one more of the heavier chlorine atoms. Many older refrigerants contained bromine as well. Table 9.4 shows some key characteristics for halogens used in refrigerants.

From the chemist's standpoint, Table 9.4 gives us some information about the nature of atomic effects on global warming. Comparing fluorine to chlorine, we can see that the mass is approximately one-half that of fluorine. The electronegativity and thus the ability to attract electrons is approximately 20% higher, and the energy required to remove 1 mol of electrons (ionization potential) is approximately 25% more difficult. These physical characteristics come into play as the inert CFCs enter the stratosphere, where they are exposed to solar radiation and provide a pathway for ozone to be converted to diatomic oxygen molecules.

Stratospheric ozone depletion by CFCs is shown in the following equations:

$$CF_2 + Cl_2 + \hbar\nu \rightarrow Cl* + CCl_2F*$$

$$Cl* + O_3 \rightarrow ClO* + O_2$$

$$ClO* + O \rightarrow Cl* + O_2$$

$$Net...Reaction: 2O_3 \rightarrow 3O_2$$

Here, the chlorine atoms act catalytically and reduce the natural balance of O_2/O_3 in the atmosphere. Each chlorine atom can react with as many as 1,000 ozone molecules.

9.8 REFRIGERANT PERFORMANCE AND SOME KEY DEFINITIONS

After the environmental effects have been considered, performance of a refrigerant system should be considered. Table 9.2 lists the normal boiling points of the most common refrigerants. The normal boiling point is critical in an HVAC system. The lower the normal boiling point is, the higher is the operating pressure of the system. Higher pressures generate greater sealing requirements as well as greater manufacturing requirements, thus generating higher cost and affecting imperatives. However, before we discuss system design, some basic definitions are in order:

Adiabatic: A process in which no heat is transferred.

Azeotropic mixture: A mixture whose components have the same composition in both the vapor and liquid phases.

Bubble point: When a binary liquid is heated, the bubble point is the point where the first bubble of vapor is formed.

Coefficient of performance (COP): A measure of an HVAC's efficiency. It is the ratio of the heat (Q) provided by the condenser to the work (W) done by the compressor; that is, $COP = |Q|/W$.

Compressor: The mechanical portion of an HVAC system that reduces the refrigerant's volume (in the vapor stage). Work is put into the system by this component.

Condenser: The component in an HVAC system that converts a refrigerant from a gas to a liquid state.

Cooling capacity: The quantity of heat in BTUs that an air conditioner or heat pump is able to remove from an enclosed space during a 1-hour period.

Critical pressure (P_{cr}): Vapor pressure at the critical temperature.

Critical temperature (T_c): The temperature above which no distinct separation between the liquid and gas phases exists.

Dew point: The temperature at which saturated vapor must be cooled at constant pressure for the vapor to condense.

Enthalpy (H): Measured in joules, $H = E + pV$; the amount of energy in a system capable of doing work; the sum of the internal energy of the system plus the product of its volume times the pressure on it. The change in enthalpy (ΔH) is equal to the heat transferred at constant pressure.

Entropy (S): Entropy has different definitions in different texts; however, it is often thought of as a measure of the randomness or disorder at a molecular level. For a system in equilibrium, $dS = \delta Q/T$ (heat absorbed for a reversible isothermal process for a state change over temperature) [8]. It can also be thought of as the unavailability of a system to do work [9]. This concept has been expanded to other fields, including information theory, psychodynamics, thermoeconomics, and evolution [10–12].

Evaporator: The component in an HVAC system where a refrigerant changes from a liquid into a gaseous state.

Heat capacity (C_p): The amount of heat energy (Q) required to raise 1 mol of a substance by 1°.

Latent heat of vaporization: Energy required to transform a quantity of substance into a gas (measured at its boiling point). It can be thought of as the energy required to overcome intermolecular interactions.

Mass flow rate: The rate at which a specific mass will move over unit time (i.e., $m = dm/dt$).

Reynolds number (Re): Ratio of inertial forces to viscous forces. Describes laminar and turbulent flow. $Re = \upsilon_s L/\nu$ (where υ_s is mean fluid velocity, L is length, and ν is kinematic fluid viscosity).

Subcooling: The temperature of a refrigerant's liquid below the boiling (condensing) temperature at a specific pressure.

Superheat: The temperature of a refrigerant's vapor above the boiling temperature of its liquid at a specific pressure.

Thermal expansion valve (TXV): The component in the HVAC system that controls the amount of refrigerant that goes into the evaporator.

Vapor compression cycle: A heating–cooling cycle that follows a refrigerant along a temperature versus entropy diagram.

Zeotropic mixture: A mixture whose components have different mass fractions in the liquid phase than in the vapor phase at equilibrium conditions.

9.9   NEED FOR ALTERNATE REFRIGERANT SYSTEMS

The European Commission has proposed legislation that will phase out any refrigerant with a global warming potential greater that 150 by May of 2008 [14]. The commission further states [14]:

> Under the Kyoto Protocol, the EU has agreed to an 8% reduction in its greenhouse gas emissions by 2008–2012, compared to the base year 1990. The reductions for each of the EU-15 countries have been agreed under the so-called EU Burden sharing agreement, which allows some countries to increase emissions, provided these are offset by reductions in other Member States.

> Emissions of the 6 greenhouse gases covered by the Protocol are weighted by their global warming potentials (GWPs) and aggregated to give total emissions in CO_2 equivalents.

> The Structural Indicator GHG emissions is [sic] defined as an index of greenhouse gas (CO_2, CH_4, N_2O, HFC, PFC and SF_6) emissions normalized by the base quantity in CO_2-equivalents (excluding land use changes and forestry). The base quantity is defined by the GHG emissions in the base year. The base year for the non-fluorinated gases (CO_2, CH_4 and N_2O) is 1990, and 1995 for the fluorinated gases (HFC, PFC and SF_6).

In the face of these changes, the automobile industry must adapt its cooling strategies to the new European requirements and possible future changes in the Americas. With the belief that satisfying the imperatives will satisfy the customer, we also address the market segment that is dedicated to the reduction of greenhouse gases and bases its purchases on environmentally friendly vehicles.

9.10   REFRIGERANT OIL MIXTURES

Automotive refrigerant systems are not composed of refrigerant alone. For lubrication purposes, an oil such as Mobil Artic EAL 68 is used. The significance of this is

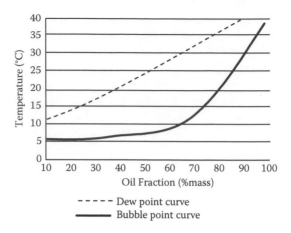

FIGURE 9.3   Phase equilibrium diagram of R134A with oil.

that, in design, the chemist must take into account how the mixture of the two liquids will behave. The mixture will be a zeotropic one in which the oil will be considered as one component even though multiple fractions are found in the oil. Figure 9.3 represents R-134a with a Mobil lubricating oil at 3.43 bar. This type of graph is called a phase equilibrium diagram. The saturation point for R134a without lubricant is 40°C. It is critical to know the bubble point and dew point when designing an evaporator for an HVAC system. Mechanical and design engineers must consult the chemist for background data.

Figure 9.4 shows another phase diagram at constant pressure. The x-axis shows the vapor–liquid mole fraction of the binary mixture. The y-axis shows temperature. The dew point line shows the temperature at which a superheated vapor mixture will begin to condense when cooled for all compositions of the mixture. The bubble point line shows the temperature at which a subcooled mixture will first begin to

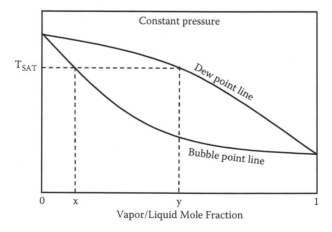

FIGURE 9.4   Vapor/liquid mole fraction diagram.

evaporate when heated. The component with the lower boiling point is represented by x. Because the refrigerants have lower boiling points, x is representative of the refrigerant; y represents the lubricating oil. If one follows the saturation temperature (T_{SAT}) across the horizontal axis, the composition of refrigerant corresponds to the composition of oil at that same temperature. The boiling point for evaporation of the zeotropic mixture corresponds to the bubble point line. The dew point line corresponds to the condensing temperature. The difference between the two is the boiling range.

9.11   152A AND HYDROCARBONS AS ALTERNATIVES

As can be seen from Table 9.2, the refrigerants 152a, CO_2, and the hydrocarbons meet the requirements for having global warming potentials below the mandated 150. However, very special care must be taken when considering R290 as a refrigerant. Table 9.5 shows some thermal and physical information for R134a, R152a, and

---

TABLE 9.5
Physical Properties of Some Refrigerants

	R134a	R152a	R290
Formula	CH_2FCF_3	CH_3CHF_2	$CH_3CH_2CH_3$
Molecular mass	102.3	66.05	44.1
Critical temperature	214	235.9	206.2
Critical pressure	589	656	616
Normal boiling point	−15	−11.2	−43.8
Lubricant	POE/PAG	NA	Mineral oil
Stability	Stable	Stable	Stable
OSHA permissible exposure limit (PPM)	1000	1000	1000
Lower flammability (% volume in air)	None	4.8	2.1
Heat of combustion (BTU/ lbm)	1806	7481	21625
Safety group	A1	A2	A3
Auto ignition temperature (°F)	1418	851	878
Atmospheric life (years)	14	2	<1
Ozone depletion potential	0	0	0
Global warming potential (100 years)[a]	1300	120	20
Latent heat of vapor at 40°F (BTU/lbm)	84.1	129.75	156.74

Source: Goodbane, M. *SAE* Paper 1999-01-084, 1999. With permission.
Notes: POE = polyolester; PAG = polyalkylene glycol; safety group = (A or B) lower and high toxicity, respectively; (1, 2, or 3) not flammable, low and high flammability.
[a]   GWP for integrated time horizon and based on 3,500 kg CO_2 per kilogram of R11.

R290. The safety group of R290 (propane) is A3, or highly flammable. Obviously, we do not want a highly flammable liquid present in a passenger compartment. This characteristic drives what would be called an "outside loop" or secondary cooling loop. In a secondary loop, the evaporator is replaced by a secondary fluid heat exchanger [7]. The system uses a direct expansion cooling system to cool a thermofluid (50% ethylene glycol/water mix), which in turn is pumped into the vehicle's heat exchanger inside the passenger compartment [7]. In addition to flammability, another obvious drawback to the R290 system is the efficiency loss with the secondary loop. Ethylene glycol has poor transportation properties and a low Reynolds number (i.e., high viscous forces).

Hydrocarbons such as R290 have been in use for years because they have good thermodynamic and transport properties, as well as low cost and low toxicity [7]. Large refrigeration plants in the petrochemical industry typically use R290. Household appliance manufacturers in Europe use isobutene (R600a) [7]. In all, U.S. auto manufacturers do not want the potential liability issues associated with the use of a hydrocarbon as a potential refrigerant [7].

For R152a, more serious consideration has been applied. The normal boiling point of R152a is slightly lower than that of 134a (Table 9.5); this means that the operating pressure of an R152a system will be slightly higher. The critical temperature of R152a is also relatively close to that of R134a, which means that the condensing temperature will be higher [7]. It can also be seen that R152a has a zero ozone depletion value (no chlorine or bromine).

Table 9.6 gives an idea of the performance of R152a as compared to R134a and R290. These data were simulated by Ghodbane. If we look at the COP, we can see a slight improvement of R152a over R134a. Over a range of temperatures, R152a can outperform R134a by 6–19% [7].

Knowledge of chemical structures and how molecules will behave based on those structures is a great asset in predicting refrigerant response. As Table 9.6 shows, the evaporator temperatures and thus much of the cooling is similar to that of R1345a without the detrimental environmental effects.

9.12 CO_2 AS AN ALTERNATIVE TO 134A

Lorentzen and Pettersen [15] sparked renewed interest in CO_2 as a refrigerant. CO_2 had been used as a refrigerant in the early twentieth century. The increased interest was in part to address the global warming issues. Brown, Yana-Motta, and Domanski evaluated the performance of CO_2 and 134a using an NIST semitheoretical vapor compression model [15]. If we compare R744 to R-134 at 5, 10, and 15°C (Table 9.7), we can see the differences in physical properties associated with the two refrigerants.

The saturation pressure (vapor pressure) of CO_2 is 10 times higher than that of R134a. Therefore, the point of boiling into its vapor phase is much higher. This translates into a system operating pressure that is 10 times higher than that of the conventional R134a system. With a pressure this high, design considerations must be taken into account. The critical pressures for CO_2 and R134a are 73.8 and 40.6 bars, respectively [16]. The critical temperature (T_c) for CO_2 is 31.1°C. Because the

TABLE 9.6
Performance of Some Typical Refrigerants at 50 mph and 40% R.H.

Parameter Description	R134a	R152a	R290
Outside air temp (°F)	100	100	100
Relative humidity (%)	40	40	40
Humidity ratio (lbm dry air)	0.017	0.017	0.017
Evap. air flow rate (CFM)	250	250	250
Evap. air mass flow rate (lbm/min)	17.3	17.3	17.3
Evap. disc. air temp (°F)	50	50	50
Condenser air flow rate (CFM)	2000	2000	2000
Condenser air mass flow rate (lbm/min)	138	138	138
Cond. air out temp. (°F)	118	117	117
Refrigerant charge (lbm)	2.125	1.375	0.918
Ref. mass flow rate (lbm/min)	6.93	4.02	3.71
Comp. suct. press (psia) = P_1	43.5	42.78	71.64
Comp. suct. temp. (°F) = T_1	38.5	40	40
Comp. disc. press. (psia) = P_2	250	224	312
Comp. disc. temp. (°F) = T_2	195	225	184
Comp. isentropic efficiency	60%	60%	60%
Comp. horsepower	4.35	3.88	4.17
Cond. out press. (psia) = P_3	238	219	297
Cond. out temp. (°F) = T_3	123	125	122
Cond. cooling capacity (Btu/min)	569	547	562
Ref. evap. in press. (psia) = P_4	53.4	47.34	82.96
Ref. evap. in temp (°F) = T_4	44	42	43
Ref. evap. out press. (psia) = P_5	48.34	45.03	77.2
Ref. evap. out temp. (°F) = T_5	38.5	40	40
Evaporator effectiveness/overall eff.	85%	85%	85%
Evap. latent load (Btu/min.)	175	175	175
Evap. sensible load (Btu/min)	207	207	207
Evap. cooling capacity (Btu/min)	382	382	382
Comp. power (BTU/min)	184	165	177
Comp. power loss (Btu/min)	5.53	4.93	5.3
A/C blower power (Btu/min)	16	16	16
System power	205	185	198
Cycle COP	2.072	2.322	2.162
System COP	1.863	2.065	1.929

Source: Goodbane, M. *SAE* Paper 1999-01-084, 1999. With permission.

TABLE 9.7
Comparison of CO_2 Refrigerant and R-134a[a]

Refrigerant	CO_2 (R-744)	R-134a
Saturation pressure (MPa)	3.69/4.502/5.086	0.35/0.414/0.488
Latent heat (kJ/kg)	214.6/196.8/176.2	194.8/190.9/186.7
Surface tension (mN/m)	3.53/2.67/1.88	11.0/10.3/9.6
Liquid density (kg/m³)	899.6/861.5/821.3	1277.1/1260.2/1242.8
Vapor density (kg/m³)	114.8/135.3/161.0	17.1/20.2/23.7
Liquid viscosity (uPa.s)	95.9/86.7/77.2	270.3/254.3/239.7
Vapor viscosity (uPa.s)	15.4/16.1/17.0	11.2/11.4/11.7
Liquid specific heat (kJ/kg K)	2.73/3.01/3.44	1.35/1.37/1.38
Vapor specific heat (kJ/kg K)	2.21/2.62/3.3	0.91/0.93/0.96

Source: Mathur, G.D. American Institue of Aeronautics and Astronautics Paper AIAA-2000–2858.
With permission.

[a] At 5°C/10°C/15°C.

T_c of CO_2 is in the ambient region, the system will operate in the transcritical mode. Because of this, the design of a system will have to incorporate a cooler instead of a condenser (see later discussion) [17].

If we examine simulated data from Table 9.8 [17], we can see that the COPs for the two refrigerants are 3.09 and 3.52 for CO_2 and R134a, respectively. Another measure of efficiency is the energy efficiency ratio, which gives the amount of cooling per unit input of electricity [17]. This efficiency is greater for 134a (value of 12) versus CO_2 (value of 9.87).

9.13   TRADITIONAL AND CO_2 REFRIGERANT SYSTEM DESIGN

Figure 9.5 shows an HVAC loop for a traditional HVAC system as well as a CO_2 system. If we follow the refrigerant path from the evaporator and move toward the compressor, refrigerant will exit the evaporator as a saturated vapor.

At stage 1, refrigerant enters the compressor (see Figure 9.6, temperature/ entropy diagram). The refrigerant gas is compressed into a superheated vapor by the compressor and then moves on to the condenser.

At stage 2, in the condenser, superheated vapor is cooled and condensed into a liquid. In this stage, heat exits the system.

At stage 3, moving out of the condenser, the refrigerant is now a saturated liquid and is routed through the thermal expansion valve (TXV).

At the thermal expansion valve (stage 4), adiabatic "flash evaporation" of part of the liquid occurs. Autorefrigeration lowers the temperature of the liquid and vapor refrigerant mixture so that it is colder than the temperature of the enclosed space to be refrigerated.

When the refrigerant returns to the evaporator (stage 5), warm air evaporates the liquid part of the cold refrigerant mixture and the cycle begins again.

TABLE 9.8
Performance Comparison of CO_2 and R-134a at Various States

State/Variables	R134a	CO_2 (R-744)
State 1		
Pressure (bar)	2.28	29.09
Temperature (°C)	−6.7	−6.7
Enthalpy (kJ/kg)	246.76	738.79
Latent heat (kJ/kg)	204.8	249.95
Entropy (kJ/kg K)	0.934078	3.99821
Volume (m³/kg)	0.088029	0.0126485
State 2		
Pressure (bar)	12.81	130
Temperature (°C)	54.4	104.4
Enthalpy (kJ/kg)	282.35	796.86
Entropy (kJ/kg K)	0.934078	3.99821
Volume (m³/kg)	0.01649	0.0041827
State 3		
Pressure (bar)	12.81	130
Temperature (°C)	48.9	26.7
Enthalpy (kJ/kg)	121.581	559.2
Entropy (kJ/kg K)	0.24211	3.230048
Volume (m³/kg)	0.0009032	0.0012006
State 4		
Pressure (bar)	2.28	29.09
Temperature (°C)	−6.7	−6.7
Enthalpy (kJ/kg)	121.581	559.2
Entropy (kJ/kg K)	0.2593667	3.288624
Volume (m³/kg)	0.000034533	0.0042816
Refrigeration effect (h1 − h4) (kJ/kg)	125.18	179.59
Cooling capacity (kW)	5.275	5.275
Ref. flow rate (kg/h)	151.58	107.2
Compressor work (kW)	1.5	1.824
Compressor capacity/volume ref. flow (kW/m³)	2318.33	19761.2
Compressor disp. of CO_2 with respect to R134a	1	8.52
Energy efficiency ratio (Btu/h W)	12	9.87
Condenser/cooler specific heat rejection (kJ/kg)	160.77	179.59
Condenser/cooler heat rejection (kW)	6.8	7.1
Coefficient of performance (COP)	3.52	3.09

Source: Mathur, G.D. American Institue of Aeronautics and Astronautics Paper AIAA-2000–2858. With permission.

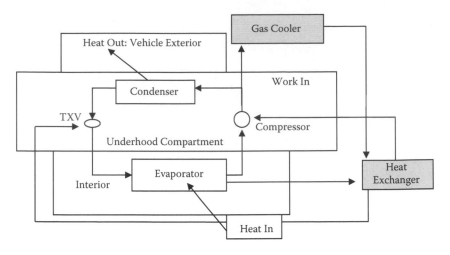

FIGURE 9.5 Diagram of traditional refrigerant and CO_2 system.

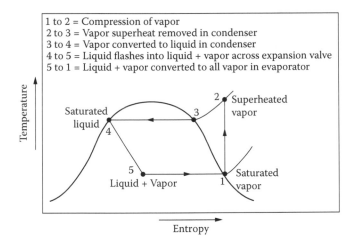

FIGURE 9.6 Vapor compression cycle.

This cycle is typically expressed in a temperature–entropy diagram called a vapor compression cycle. Figure 9.6 illustrates this cycle.

A CO_2 system operates slightly differently and requires an outside loop. The deviation is because CO_2 operates in the transcritical mode at typical ambient conditions (as mentioned earlier) and requires much higher pressures than a typical refrigerant system. Above the critical temperature (T_c), distinct liquid and gas phases do not exist. Properties between the two phases are the same as one approaches this temperature. Above T_c, a liquid cannot be formed; however, a solid can be formed with enough pressure. The T_c for CO_2 is 31.1°C, whereas a typical system with R134a refrigerant is 101°C. In automotive applications, condensing in a refrigeration cycle takes place at around 38–60°C. Because of this, the condenser used in typical applications is replaced by a gas cooler that rejects heat.

TABLE 9.9
CO_2 Effects on Average Adults

CO_2 in Air (%)	Effect on Average Adult
2	50% increase in breathing rate
3	100% increase in breather rate; 10 min short-term exposure limit
5	300% increase in breathing rate; headache and sweating may begin after 1 h; this is tolerated by most persons but is physical burdening
8	Short-term exposure limit
8–10	Headache after 10–15 min; dizziness; buzzing in ears; rise in blood pressure; high pulse rate; excitation; nausea
10–18	Cramps after a few minutes; epileptic fits; loss of consciousness; a sharp drop in blood pressure (victims recover quickly in fresh air)
18–20	Symptoms similar to those of a stroke

In addition, an internal heat exchanger is needed to improve performance. This component will subcool the refrigerant between the TXV and gas cooler. The heat exchanger will also superheat the refrigerant in the suction line of the compressor. This secondary loop system is on outside the vehicle compartment (as shown in diagram in Figure 9.1) to assure protection against leaks. Secondary loop systems are also used in hydrocarbon systems as well as 152a refrigerant systems for flammability reasons. Table 9.9 demonstrates the effects of CO_2 on a human adult.

9.14 NEW DEVELOPMENTS IN REFRIGERANT DESIGN (1234YF)

Another likely candidate to replace R134a is HFO-1234yf, which was developed by Honeywell and DuPont. This new refrigerant was driven by the European Union's F-gas phase-out of 134a for new vehicle models. R-134a will be banned by January 2017. New refrigerants will have GWPs of less than 150. HFO-1234yf has a GWP of 4, well under the limit of 150 set by the EU. The components of HFO-1234yf are hydrofluoro-olefin with a secondary ingredient of trifluoro(iodo)methane. This is shown in Figure 9.7. The secondary ingredient, trifluoro(ido)methane, is a cardiac sensitizer, which causes concerns about the safety of the refrigerant.

In addition, HFO-1234yf is slightly flammable, but can potentially be used without a secondary loop. This would be an advantage over a CO_2 system, which would most likely require a secondary loop. HFO-1234yf is flammable in concentrations

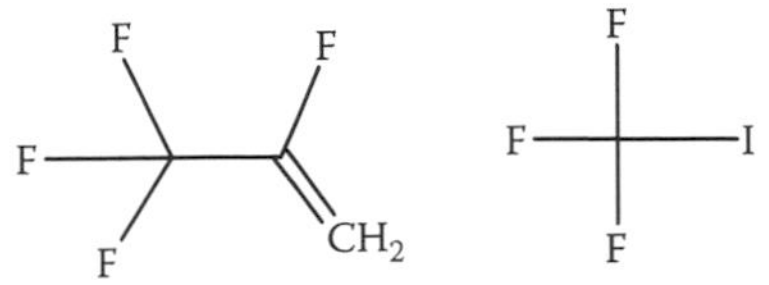

HFO-1234yf hydrofluoro-olefin

FIGURE 9.7 Structure of HFO-1234yf.

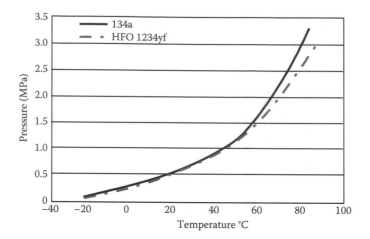

FIGURE 9.8   Pressure–temperature curve of 134a versus HFO-1234yf.

of 6.5–12.3% in air, as compared to R-152a, which is 3.9–16.9%. In recent drop-in-replacement tests, HFO-1234yf had slightly better cooling performance than R134a. Figure 9.8 shows the pressure temperature curve between HFO-1234yf and R-134a. The curve is practically line on line until about 60°C.

If we look at a properties chart (Table 9.10), we can see that critical performance characteristics such as boiling point, critical point, and vaporization pressures are nearly the same. The major difference is the GWP of 4 as compared to the value of 1,300 for 134a.

Another great advantage is that the extreme pressure requirements required by a CO_2 system are not required for the drop in HFO-1234a. Work is ongoing on HFO-1234yf systems as safety standards, including toxicity, flammability, and others that must be met in order to be accepted by the industry. Overall, the implementation of HFO-1234yf looks very promising; tests indicate that cooling capacity and energy efficiency numbers generated are within 5% of the current R-134a values.

TABLE 9.10
Properties of HFO-1234yf and 134a

Properties	HFO-1234yf	134a
Boiling point	−29°C	−26°C
Critical point, T_c	95°C	102°C
P_{vap} MPa (25°C)	0.677	0.665
P_{vap} MPa (80°C)	2.44	2.63
Liquid density (kg/m³)	1094	1207
Vapor density (kg/m³)	37.6	32.4
GWP	4	1300

TABLE 9.11
HVAC System Components and Materials

Component	Typical material used
HVAC module case	Polypropylene (G/F)/nylon 6,6
HVAC module sealer foam	Polyurethane open/closed cell foam
Valves	Polypropylene (mica filled/glass filled/mineral filled)
Heater core	Aluminum tube AA4343/AA3003 base
Evaporator core	Aluminum tube AA4343/AA3003 base
Electric cooling fan	Nylon 6,6/nylon 6/polypropylene
Engine-driven fan	Nylon 6,6/nylon 6
Engine-driven fan shroud	Polypropylene (GF)
Hoses	EPDM
AC lines	Aluminum tube AA4343/AA3003 base
Condenser	Aluminum tube AA4343/AA3003 base
Radiator	Aluminum tube AA4343/AA3003 base
Isolators	EPDM or natural rubber (steel/nylon inserts)
Charge air coolers	Aluminum AA4343/AA3003/nylon

9.15 MATERIAL CONSIDERATIONS IN HVAC DESIGN

As with any automotive design, the design engineer must consult the chemist (i.e., material engineer) about the requirements for proper functioning of the parts. Typical inputs will be temperature, pressure, friction, and chemical environment. As we have discussed, some of the components that make up a typical list can be found in Table 9.11.

Material selection is extremely important, not only because of system requirements (i.e., temperatures, pressures, etc.) but also because of material cost considerations. In an HVAC module, the heater core and evaporator core reside within the module, which usually comprises nylon 66 or a filled polypropylene. Refrigerant flows through the evaporator core and condenser and will vary in a range from 9 to 115°C. Engine coolant flows through the heater core and reaches temperatures up to 120°C. Valves and seals within the HVAC module must withstand temperature cycling combined with pressure extremes.

A typical module contains a mode valve, air-inlet valve, and temperature valve. The mode valve controls the various modes of the AC unit (i.e., heater, defrost, AC, blend, and bi-level). Pressure is asserted on this valve at the varying temperatures due to airflow coming from the heater and evaporator core. The temperature valve is basically a mix valve between the heater core and evaporator core and as such must endure the most extreme temperatures. Typical properties for the plastic portion used in HVAC systems are shown in Table 9.12.

TABLE 9.12
Polypropylene and Polyamide Properties

	Polypropylene	Polyamide
	Glass fiber reinforced, chemically coupled, heat stabilized homopolymer	Glass fiber reinforced, mineral filled, heat stabilized
Heat deflection temperature at 1820 kPa	141°C	205°C
Ultimate tensile strength	69 MPa	93 MPa
Flexural modulus	4140 MPa	6000 MPa
Glass content	30 ± 2%	35–42%

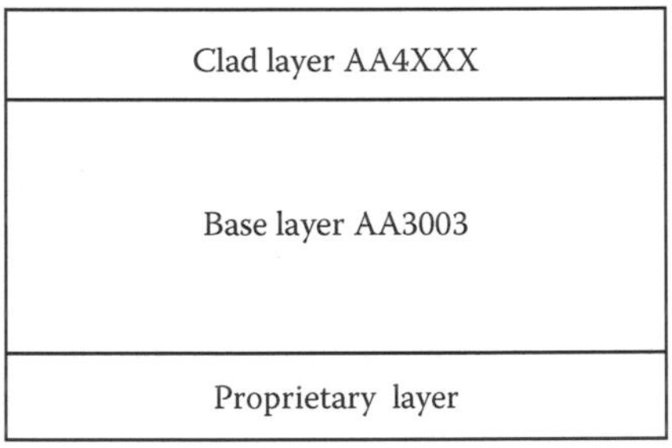

Heat Exchanger Tube Material
with Cladding for Brazement

FIGURE 9.9 Material cross section of heat exchanger material.

9.16 ALUMINUM HEAT EXCHANGER MATERIAL

In heat exchanger design, the most typical material used for tubes and air conditioning lines is Aluminum Association 3003. Most suppliers will have this base layer modified with a clad layer (typically, AA4XXX series) on the exterior for brazing operations. The principal design feature for this is its corrosion resistance and thermal properties. For erosion resistance, a proprietary third layer is added to the interior of a tube. Figure 9.9 shows a schematic for heat exchanger tubes.

REFERENCES

1. Molina, M., and F. Rowland. 1974. *Nature* 249:810.
2. Scitech Instrument, UV sensor application note: UV index measurement. February 26, 2008. http://www.scitech.uk.com
3. http://www.theozonehole.com/montreal.htm
4. Use of ozone depleting substances in laboratories. TemaNord 2003:516. http://www.norden.org/pub/ebook/2003-516.pdf

4a. Farman, J. C., B. G. Gardiner, and J. D. Shanklin. 1985. Large losses of total ozone in Antarctica reveal seasonal $C10_x/NO_x$ interaction. *Nature* 315.

5. http://www.ccpo.odu.edu/SEES/ozone/class/Chap_10/10_4.htm

6. http://www.grida.no/climate/ipcc_tar/wg1/247.htm

7. Ghodbane, M. 1999. An investigation of R152a and hydrocarbon refrigerants in Mobil air conditioning. SAE paper 1999-01-0874, pp.1, 2, 4, 10, 14.

8. Perrot, P. 1998. *A to Z of thermodynamics.* Oxford, England: Oxford University Press.

9. Dainth, J. 2005. *Oxford dictionary of physics.* Oxford, England: Oxford University Press.

10. Avery, J. 2003. *Information theory and evolution.* Singapore: World Scientific Publishing.

11. Yockey, H. P. 2005. *Information theory, evolution, and the origin of life.* Cambridge University Press.

12. Brooks, D. R., and E. O. Wiley. 1998. *Entropy as evolution—Towards a unified theory of biology.* Chicago: University of Chicago Press.

13. http://www.energystar.gov/ia/partners/prod_development/archives/downloads/lchvac/Draft_LC_HVAC.pdf

14. http://europa.eu.int/estatref/info/sdds/en/env/env_air_sm.htm

15. Lorentzen, G., and J. Petterson. 1992. New possibilities for non-CFC refrigeration. In *Proceedings of the IIR International Symposium on Refrigeration, Energy, and Environment,* Trondheim, Norway, 147–163.

16. Brown J., S. Yana-Motta, and P. Domanski. 2002. Comparative analysis of an automotive air conditioning system operating with CO_2 and R134a. *International Journal of Refrigeration* 25:374.

17. Mathur, G. D. 2000. *Carbon dioxide as an alternative refrigerant for automotive air conditioning systems,* 371, 376. American Institute of Aeronautics and Astronautics AIAA-2000-2858. Energy Conservation Engineering Conference and Exhibit.

10 Fuel-Cell Chemistry Overview

10.1 INTRODUCTION

Companies such as Daimler, Ford, GM, Honda, Peugeot–Citroen, Toyota, and Renault–Nissan have long known the limitations of internal combustion engines. The efficiency of petroleum-fueled engines is expected to peak at around 30% [1]. Efficiency is not the only limitation for internal combustion engines. As detailed in Chapter 9, combustion of gasoline produces emissions that are environmentally harmful. Hydrocarbons, CO_2 emissions, nitrogen oxides, and carbon monoxide are released into the atmosphere during combustion. Predictions are that fuel-cell vehicles will be twice as efficient as internal combustion engines and have only by-products of water and heat [1].

With advances in technology such as direct fuel injection, improvements in EGR valves, and improvement in cylinder sealing, internal combustion engines have greatly improved in efficiency. Since the 1960s, CO_2, hydrocarbon emissions, and nitrogen oxides have decreased by over 90% [1]. Despite these improvements, CO_2 emissions remain a cause for concern due to their global warming potential (discussed in Chapter 9). One of the benefits of fuel-cell vehicles is that they produce no polluting emissions and utilize a renewable source of energy.

General Motors and Toyota are also aggressively pursuing battery-operated vehicles; however, these have certain drawbacks as well. Battery-powered vehicles only store a finite amount of energy. Range for these vehicles is limited typically to about 100 miles, and cooling of the lithium ion battery is currently an issue [1]. The hybrid vehicles currently being produced are a great help with fuel efficiency and environmental issues are only a stop-gap measure. With these technologies, a fundamental paradigm shift in automotive as well as world thinking around energy and its efficient use is currently occurring.

10.2 FUTURE MARKET AND USAGE

Recent studies by the WBCSD (World Business Council for Sustainable Development) have been published and some very interesting findings noted [2]. For instance, with a current population of motor vehicles of approximately 600 million, the WBCSD points out that the world population is urbanizing; it projects 8 billion people living in urban areas by 2030 and fuel use to be 3.8 trillion L by 2030 [2]. If population predictions are accurate, then as many as 1.2 billion persons will have vehicles by 2030 [3]. As countries like China and India develop their middle class, we can expect

TABLE 10.1

Personal Transportation Activity by Region—Average Annual Growth Rates

	2000–2030 (%)	2000–2050 (%)
Africa	1.90	2.10
Latin America	2.80	2.90
Middle East	1.90	1.80
India	2.10	2.30
Other Asia	1.70	1.90
China	3.00	3.00
Eastern Europe	1.60	1.80
Former Soviet Union	2.20	2.00
OECD Pacific	0.70	70.00
OECD Europe	1.00	80.00
OECD North America	1.20	1.10
Total	1.60	1.50

Source: Sustainable Mobility Project calculation. http://www.wbcsd.org/tem-
plates/Template WBCSD5/Layout.asp?MenuID=1

that the corresponding increase in income will be followed by a demand for personal vehicles. Table 10.1 shows the average annual growth rate as predicted by the Sustainable Mobility Project, which is part of the WBCSD [2].

10.3 FUEL CELLS AS AUTOMOTIVE PROPULSION

There are many types of different fuel cells; however, for our purposes we will focus on proton exchange membrane (PEM) fuel cells because they are most applicable to automobiles. A PEM basically consists of an anode and cathode in a thin cell separated by a polymer membrane. There is a platinum catalyst, which is electrochemically active. Also, of course, hydrogen and oxygen are in the circuit. Figure 10.1 illustrates the process of how fuel cells work.

Hydrogen enters the fuel cell from the source and is exposed to the anode, where an electron is extracted, leaving a proton via

$$H_2 \rightarrow 2H^+ + 2e^- \tag{10.1}$$

The electrons are now free to go to work in the vehicle by driving an electric motor. Stripped of their electrons, the protons travel across a membrane of perfluorosulfonic acid (Nafion by DuPont). Proton conductivity is enhanced by impregnation with an acidic solid. The resulting solution will provide a higher concentration of protons. The catalyst on the cathode interface will combine the protons with oxygen from the air, producing a potential and causing the electrons to move through the

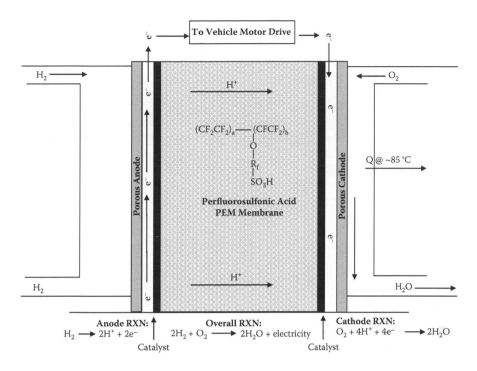

To Vehicle Motor Drive

H_2

O_2

H^+

$(CF_2CF_2)_a$ — $(CFCF_2)_b$

O

R_f

SO_3H

**Perfluorosulfonic Acid
PEM Membrane**

Porous Anode

Porous Cathode

$Q @ \sim 85\,°C$

H^+

H_2

H_2O

Anode RXN:
$H_2 \longrightarrow 2H^+ + 2e^-$
Catalyst

Overall RXN:
$2H_2 + O_2 \longrightarrow 2H_2O + electricity$

Cathode RXN:
$O_2 + 4H^+ + 4e^- \longrightarrow 2H_2O$
Catalyst

FIGURE 10.1   Diagram of fuel cell.

external circuit. As a result of the current through the circuit, the oxygen will receive two electrons and become negatively charged [4]. The cathode reaction is shown in Equation 10.2:

$$O_2 + 4H^+ + 4e^- \rightarrow 2H_2O \qquad (10.2)$$

Because positively charged protons are on one side of the membrane and negatively charged oxygen ions are on the other, diffusion across the membrane will occur [4]. At the cathode, water is formed and the overall reaction is

$$2H_2 + O_2 \rightarrow 2H_2O + electricity \qquad (10.3)$$

PEM-type fuel cells have the advantage of operating at pressures from 101 to 810 kPa. In addition, they work well at lower temperatures (a feature not present in other types of fuel cells). These fuel cells can operate at 55% efficiency versus a maximum of 30% for internal combustion engines [1]. In order to achieve the necessary energy to power the motors for propulsion, fuel cells are manufactured in a stack arrangement. In these arrangements, other components are needed (such as humidifiers to keep the cell moist, air compressors, and gas filters) to make up the fuel-cell propulsion unit. Each stack has a flat cathode sheet and a flat anode sheet, which are stacked in order to meet the particular vehicle's power requirements [1]. Current vehicle testing by Toyota and Daimler indicates power ranges from 25 to 50 kW (34–68 hp); other manufacturers plan on methanol cells with up to 65 kW (88 hp) of power [4].

10.4   HYDROGEN SOURCES

Currently, reforming hydrogen molecules with some sort of catalyst is being pursued by many of the automotive companies. Methanol and gasoline reformation seems to be the preferred method. Basically, hydrogen reforming is a method of producing hydrogen from hydrocarbons. In large-scale, inexpensive, industrial reforming, SMR (steam methane reforming) is used [5]. This method uses nickel as a catalyst with methane to produce CO and H_2 (Equation 10.4). This reaction takes place from 700 to 1100°C:

$$CH_4 + H_2O \rightarrow CO + 3H_2 \qquad\qquad (10.4)$$

Proposed fuel-cell vehicle designs utilize a hydrogen tank, which is quite large, bulky, and potentially dangerous. If an alcohol tank is used with steam reformation, however, it may be feasible to eliminate the associated problems with the pressurized tank system. In a green economy, the ideal fuel source would be a renewable source of fuel such as biomass.

Using alcohol and gasoline for a source of hydrogen can be thought of as an interim solution until the problem of adequate hydrogen storage can be solved. Again, the chemist's role is critical in the development and execution of this process. One advantage of this interim solution is that the current infrastructure of gasoline stations could be utilized. Fuel reformers could be installed at these stations, which could allow for hydrogen to be provided at the station [1].

One plan that has been suggested is that some automobile companies provide fuel-cell generators capable of 75 kW of power to be sold to hospitals and the like [1]. The advantage of a business decision such as this would be threefold: Valuable knowledge and experience could be gained from the manufacture and production of these units. Reliability data could be gathered, manufacturing improvements made, and trust built for the manufacturer. In addition, income could be generated from the sale of these generators. This income could support investments made into the technology. A third advantage would be that generated power could be sold back to the grid during peak periods [1].

Bruns, McCormick, and Borroni-Bird addressed the cost of reformed hydrogen versus gasoline [1]. They stated that the cost of a kilogram of hydrogen could be four to six times as high as the cost of a gallon of gasoline or diesel. However, due to the increased efficiency of the fuel cell, 1 kg of hydrogen would be equivalent in cost to 1.3 gal of traditional fuel [1].

10.5   PROBLEMS WITH FUEL CELLS

As with any new technology, problems exist that must be overcome in order for that technology to work. Obviously, even though the idea of fuel cells has been around for quite some time, implementation of the cells has not taken place for various reasons. These reasons are being addressed by OEM (original equipment manufacturer) chemists in labs today, keeping in mind the imperatives of automobile design.

10.5.1 OVERPOTENTIAL

Energy is lost due to overpotential at the anode and cathode. A certain amount of voltage is needed to overcome the potential barrier of oxygen and hydrogen at the electrode reactions [6]. Chemists are currently working on this problem by looking for more efficient materials. A steep voltage drop occurs at high current density due to mass transfer of the reacting species at the electrode. Svoboda and colleagues report densities of 0.8–1.2 amps/cm^2 for a single cell for a range of 0.55–0.75 V [6].

10.5.2 TEMPERATURE CONSIDERATIONS

Reforming reaction takes place at high temperatures. Typically, in vehicle systems utilizing methanol reformation, hydrogen is directly injected into the cell where methanol and water are vaporized to form H_2, CO, and CO_2 [6]. This reformation takes place at approximately 280°C [6]. Materials that operate at these temperatures are quite costly. In addition, separate loop cooling systems (currently in the form of humidifiers) are required to keep the overall cell temperature down and the water management under control for a PEM fuel cell. An aqueous membrane is required; to prevent boiling, temperatures below 100°C are necessary [6].

10.5.3 SULFUR COMPOUNDS

Sulfur compounds present in the fuel will contaminate certain catalysts, making it difficult to run this type of system from ordinary gasoline. Work being done by Kataria, Ayyappan, and Abraham around rhodium-based, sulfur-tolerant catalysts at 800°C with jet fuel may be pertinent to automotive applications [7].

10.5.4 CARBON MONOXIDE

Carbon monoxide produced by the reactor will poison the fuel cell. This makes it necessary to include complex CO removal systems [8]. Nafion membranes are very sensitive to CO. The process of chemisorption (adherence to the surface via a chemical reaction) takes place, with CO competing with H_2 on the platinum surface [4].

10.5.5 CATALYST COST

Catalysts that are currently being used are made of platinum or rhodium. These metals are very expensive. According to the "Platinum Metals Report" [9], platinum prices in 2008 were on the rise again due to uncertainty over power supply interruptions in South Africa (see Figure 10.2). The price of the platinum metal was at $2,180 per ounce. The record rapid rise shows the volatility and sensitivity of the market and

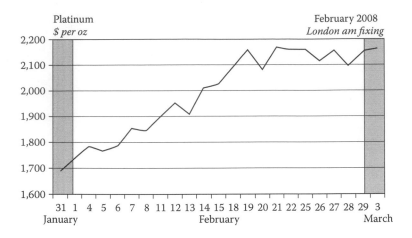

FIGURE 10.2 Chart of platinum prices for 2008.

how it can be affected by such things as power supply in mining operations. Prices are at an all-time high.

Even worse is the price of rhodium. According to Maycock (seekingalpha.com [10]), the two metals are "on fire." Rhodium was at $8,500 per ounce due to the same issues affecting platinum. Unfortunately, most research for a sulfur-tolerant catalyst has been with rhodium. Figure 10.3 shows a 2008 chart of rhodium prices from platinum.matthey.com [11]. The chart shows a period average of $7,760, with a peak of over $9,660 in June of 2008. Many factors affect these prices, such as strikes, global markets, and energy concerns. The key lesson is that innovation should drive chemists and engineers to create designs that are insensitive to the noise of the markets.

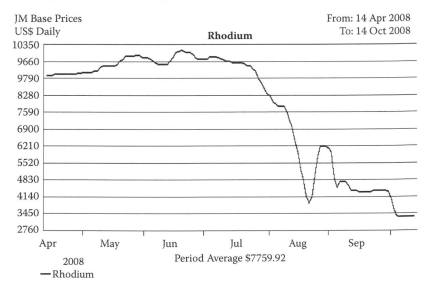

FIGURE 10.3 Chart of rhodium prices.

10.5.6 Hydrogen Storage

Hydrogen storage is a formidable problem being worked on by chemists and engineers. We will discuss this in more detail in Chapter 11.

10.5.7 Vehicle Design

Much speculation has taken place around the design of fuel-cell vehicles and how radical changes can take place with differing architecture. For instance, in *Scientific American,* GM discusses its autonomy concept and a drivable prototype called "Hy-wire" [1]. A rough schematic of this concept is presented in Figure 10.4.

The General Motors concept would incorporate what is known as "drive by wire," which uses electrical devices to supplant the use of mechanical linkages within a vehicle. The basic concept design is a ground-up design utilizing fuel cells [12]. The chassis design was rolled out in 2002 at the Detroit Auto Show and was 11 in. thick [12]. In addition to the fuel-cell chemistry, opportunity and, indeed, chemist input are required. Research on material choices that comprise the chassis (i.e., engineering polymer choices), interior components, coolant choices for the heat exchangers, and crash-absorbent materials in the crush zone is needed. The chemist is best equipped to do this research. Some GM engineers have speculated that with this concept, there could be interchangeable chassis and only one single connection to the set of mechanical links that unite the concept [1]. In this type of system, functions such as braking, throttle control, and steering are controlled through a computer. Electronic throttle control (ETC) directly links the accelerator pedal and the throttle. Instead of using a TPS (throttle position sensor), which provides input to antilock brakes, fuel injection, and traction control, and data through a cable, a solenoid is utilized to control the throttle [13].

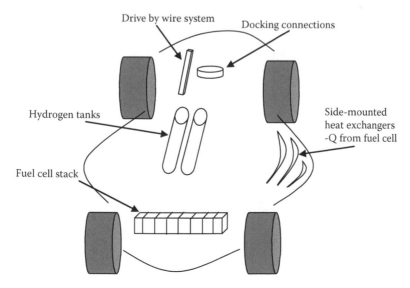

FIGURE 10.4 Diagram of potential fuel cell vehicle.

One of the benefits of driving by wire is the ability to change vehicle design radically, as with the GM autonomy concept. The chassis is essentially a skateboard or flat chassis that lowers the CG (center of gravity) as well as permits radical design concepts and the ability to interchange chassis. In addition, computer upgrades can be made to switch from, say, a family vehicle to a sportier chassis [1]. An electronic throttle control using solenoids can increase the fuel delivery and acceleration of a vehicle as well as improve its fuel economy [13].

In 2005, Toyota introduced the Prius. This vehicle utilizes drive-by-wire technology for throttle, brake, and transmission control. The engine control is done by the "hybrid synergy drive system," which assigns control to a hybrid control computer [13]. Most of the technology currently exists for implementation of fuel-cell vehicles, and many of the concepts are already in production. Some infrastructure already exists along the Gulf Coast of the United States, Europe, and the Netherlands [1]. Hydrogen used for removal of sulfur in petroleum refining is already at approximately 540 billion m^3 [1]. However, with onboard hydrogen tanks, perceived safety is an issue. The problem of hydrogen storage needs to be addressed and agreed upon. Some promising work has been done in this field and will be discussed in the next chapter.

REFERENCES

1. Bruns, L., B. J. McCormick, and C. E. Borroni-Bird. 2002. How fuel-cell cars could revolutionize the world. *Scientific American* October: 4–7, 11.
2. http://www.wbcsd.org/web/publications/mobility/overview.pdf
3. http://hypertextbook.com/facts/2001/MarinaStasenko.shtml
4. http://www.cem.msu.edu/ cem181h/projects/97/fuelcell/chem1.htm
5. Crabtree, G. W., M. S. Dresselhaus, and M. V. Buchanan. 2004. The hydrogen economy. *Physics Today* 57 (12): 39.
6. http://www.cem.msu.edu/cem181h/projects/97/fuelcell/chem1.htm
7. http://aichi.confex.com/aiche/2007/preliminaryprogram/abstract_85948.htm
8. http://en.wikipedia.org/widi/steam_reforming
9. http://www.platinum.matthey.com/uploaded_files/prices0208.pdf
10. http://seekingalpha.com/article/65134-platinum-and-rhodium-two-metals-on-fire
11. http://www.platinum.matthey.com/prices/price_charts.html
12. http://www.popularmechanics.com/automotive/new_cars/1266806.html
13. http://wikicars.org/en/Drive-by-wire_throttle

11 Membranes and Hydrogen Storage Devices

11.1 INTRODUCTION

As discussed in Chapter 10, fuel-cell vehicles face the problem of hydrogen storage. Consumers perceive an onboard storage tank as a safety issue. However, a much larger issue is tank size. Hydrogen can be stored as a compressed gas, in a rechargeable metal hydride, in a hydride compound that releases hydrogen when reacted with water, through a cooled liquid, or through adsorption on a surface [1].

- Proton exchange membrane (PEM) fuel cells are the best choice for automobiles due to their high-power density, low mass, and low operating temperatures as compared to other types of fuel cells.
- Alkaline fuel cells (AFCs) have been around the longest amount of time. They utilize potassium hydroxide and are susceptible to carbon contamination.
- Phosphoric acid fuel cells (PAFCs) utilize liquid phosphoric acid as the electrolyte. These types of cells can achieve efficiencies up to 80%. They are usually large and heavy and require warm-up time.
- Molten carbonate fuel cells (MCFCs) use an electrolyte composed of a molten carbonate salt mixture suspended in a porous, chemically inert ceramic lithium aluminum oxide ($LiAlO_2$) matrix. MCFCs are large and operate at very high temperatures (1,200°F). Because they use a corrosive electrolyte, their durability is limited.
- Solid oxide fuel cells (SOFCs) use a nonporous ceramic compound as the electrolyte and operate at very high temperatures (1,800°F). Heat can be recaptured for co-generation, making these fuel cells highly efficient (80–85%). Because of size, heat output, and a long start-up time, these fuel cells are more suitable for stationary applications.

11.2 HYDROGEN STORAGE TANK SIZE

Compressed hydrogen tanks that are currently utilized require 6.8 kg of hydrogen to power a 1,500-kg (3,307-lb) vehicle [1]. At this mass, a 340-L tank (at 25 MPa) would be required for a 350-mile range [1]. A mass of 3,307 lb is the approximate average size of an American passenger car. With the high prevalence of SUVs in the American market, an even larger tank would be required. These vehicles have masses in excess of 5,000 lb and towing requirements in excess of 11,000 lb. A 70-L gasoline tank would be required to store the amount of fuel necessary for the previous example. In Table 11.1, the "fuel type" column shows what is offered today for hydrogen storage. As can been seen, compressed hydrogen or some form of hydrogen that needs re-formation is being utilized.

11.3   NEW DEVELOPMENTS

Automotive manufactures such as Daimler–Benz and Toyota are utilizing methanol and gasoline reformation as their main strategies in fuel cell development. As mentioned earlier, methanol and water vaporize to form carbon monoxide, carbon dioxide, and hydrogen. Direct storage of hydrogen in an elemental state would be beneficial by shrinking the large size of the storage tanks as well as eliminating the necessity for re-formation.

11.4   GLASS MICROSPHERES

Glass microspheres—glass spheres usually ranging from 1 to 1,000 µm [2]—have been thought of as one possibility for hydrogen storage in fuel cells. Hollow microspheres have diameters up to 300 µm [3]. Glass microspheres have been used in other applications, such as sieve and filter calibration, embolization therapy for capillary blockage, and slow release of pharmaceuticals. Hollow microspheres can be used as lightweight filler in composites and lightweight concrete [4]. They can also be used for storage and slow release of pharmaceuticals [4].

Glass microspheres are made by heating tiny droplets of dissolved water glass in a process known as ultrasonic spray pyrolysis, which treats the glass with acid to remove sodium [4]. This process is generally used in the production of nanoparticles [5]. Glass microsphere thicknesses are generally about 1 µm. When the spheres are heated to 200–400°C, permeability of the glass is possible [6]. Hydrogen gas can then be filled into the void of the hollow tube. Temperatures are then lowered and the hydrogen is trapped inside. Inside a fuel cell vehicle, the spheres can be heated to release hydrogen.

11.5   CARBON NANOTUBES AND GRAPHITE NANOFIBERS

Carbon in graphite form arranges itself in honeycomb sheets. Figure 11.1 shows a typical representation. A nanotube is a set of graphite sheets wrapped around one another that create a very strong material. Graphite is made as a by product of a catalytic reaction. It was discovered by Terry Baker while he was at the Atomic Energy Authority in Harwell, England [6]. At that time, the fibers were considered a nuisance material. As a catalytic reaction proceeds, carbon is stacked above and below the metal particle that is used. The arrangement of the platelet is dependent on the catalytic metal used.

One of the most important factors of these fibers is the consistent distance between each fiber. Figure 11.2 shows the length and diameter of the carbon nanotubes: 5–100 µm long and 5–100 nm in diameter. The configuration is actually tubular in construction. The ends of the tube can be opened by being treated with nitric acid and the resulting opening of 5–100 nm gives the tube its name. The atomic radius of

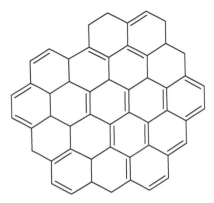

FIGURE 11.1   Typical representation of carbon in graphite form.

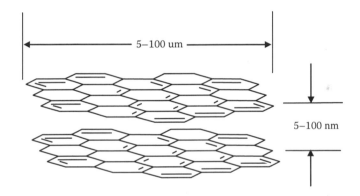

FIGURE 11.2   Diameters of carbon nanotubes.

hydrogen is 0.35 Å; an H_2 molecule is only slightly larger than 2.6 Å and thus is just large enough to take up residence within these nanotubes [1].

The amount of hydrogen that can be stored on these tubes has been debated. One report shows that maximum storage capacity for a single-walled carbon nanotube (SWCNT) [7] and multiwalled carbon nanotubes [8] is approximately 8% by weight [9]. Multiwalled carbon nanotubes are a collection of concentric single-walled nanotubes [10] versus the one-dimensional single-walled tubes [11]. SWCNTs have very high surface-to-volume ratios as well as uniform pores, which allow for capillary action and thus the ability to be filled by condensation.

Adsorption on the interior surface of the tubes can take place. At high temperatures, hydrogen gas condenses to a liquid deep within the tubes, squeezing between the carbon nanotube plates and thus decreasing their vibration activity; this allows for greater compression [1]. To get the hydrogen into the fuel cell, a vacuum is pulled and the tube heated to 900°C. When hydrogen is inserted into the tube, a pressure of 12,000 kPa is used [1]. A pressure of 4,000 kPa must be maintained to keep the H_2 in place in the cell [1].

TABLE 11.1
Fuel Cell Manufacturers and Vehicle Data

Automaker	Vehicle Type	Engine Type	Fuel Cell Size/Type	Fuel Cell Manufacturer	Range (mi/km)	Maximum Speed	Fuel Type
Audi	A2	Fuel-cell/battery hybrid	66 kW/PEM	Ballard	220 km	175 km/h	Gaseous hydrogen
BMW	Series (745 h) (Sedan)	ICE (fuel-cell APU)	5 kW/PEM	UTC	180 mi/300 km	140 mph	Gasoline/liquid hydrogen
Daihatsu	Move FCV-K II (minivehicle)	Fuel-cell/battery hybrid	30 kW/PEM	Toyota	75 mi/120 km	65 mph/105 km/h	Compress hydrogen at 3600 psi
Daimler–Chrysler	NECAR 1 (180 van)	12 fuel cell stacks	50 kW/PEM	Ballard	81 mi/130 km	56 mph/90 km/h	Compress hydrogen at 4300 psi
	NECAR 2 (V-Class)	Fuel cell	50 kW/PEM	Ballard	155 mi/250 km	68 mph/110 km/h	Compress hydrogen at 3600 psi
	NECAR 3 (A-Class)	Two fuel-cell stacks	50 kW/PEM	Ballard Mark 700 series	250 mi/400 km	75 mph/120 km/h	10.5 gal of liquid methanol
	NECAR 4 (A-Class)	Fuel cell	70 kW/PEM	Ballard Mark 900 series	280 mi/400 km	90 mph/145 km/h	Liquid hydrogen
	Jeep Commander 2 (SUV)	Fuel-cell/ (90 kW) battery hybrid	50 kW/PEM	Ballard Mark 700 series	118 mi/190 km	N/A	Methanol
	NECAR 4 Advanced (California NECAR)	Fuel cell	85 kW/PEM	Ballard Mark 900 series	124 mi/200 km	90 mph/145 km/h	4 lb of compressed hydrogen at 5000 psi
	NECAR 5 (A-class)	Fuel cell	85 kW/PEM	Ballard Mark 900 series	280 mi/450 km	95 mph/150 km/h	Methanol

NECAR 5.2 (A-class)	Fuel-cell/battery hybrid	85 kW/PEM	Ballard Mark 900 series	300 mi/482 km	95 mph/150 km/h	Methanol
Springer (van)	Fuel cell	85 kW/PEM	Ballard Mark 900 series	93 mi/150 km	75 mph/120 km/h	Compress hydrogen at 5000 psi
Natrium (Town and Country minivan)	Fuel-cell/(40 kW) battery hybrid	54 kW/PEM	Ballard Mark 900 series	300 mi/483 km	80 mph/129 km/h	Catalyzed chemical hydride-sodium Borohydride
F-cell (A-class)	Fuel-cell/battery hybrid	85 kW/PEM	Ballard Mark 900 series	90 mi/145 km	87 mpg/140 km/h	4 lb of compressed hydrogen at 5000 psi
Jeep Treo	Fuel cell	N/A	N/A	N/A	N/A	N/A
F600 HYGENIUS	Fuel-cell/battery hybrid	60 kW/PEM	N/A	250 mi/400 km	105 mph	Hydrogen
EcoVoyater	Fuel-cell/battery hybrid	45 kW/PEM		300+ mi	115 mph	Compressed hydrogen at 10,000 psi
ESORO Hycar	Fuel-cell/battery hybrid	6.4 kW/PEM	Nuvera	224 mi/360 km	75 mph/120 km/h	Compressed hydrogen
Fiat Seicento Elettra H2 fuel cell	Fuel-cell/battery hybrid	7 kW/PEM	Nuvera	100 mi/140 km	60 mph/100 km/h	Compressed hydrogen
Seicento Elettra H2 fuel cell	Fuel-cell/battery hybrid	N/A	Nuvera	N/A	N/A	Compressed hydrogen
Panda	Fuel cell	60 kW/PEM	Nuvera Andromeda II stack	125 mi/200 km/h	81 mph/130 km/h	Compressed hydrogen
Ford Motor Company P2000 HFC (sedan)	Fuel cell	75 kW/PEM	Ballard Mark 700 series	100 mi/160 km	N/A	Compressed hydrogen
Focus FCV	Fuel cell	85 kW/PEM	Ballard Mark 900 series	100 mi/160 km	80 mph/128 km/h	Compressed hydrogen at 3600 psi

—continued

TABLE 11.1 (continued)
Fuel Cell Manufacturers and Vehicle Data

Automaker	Vehicle Type	Engine Type	Fuel Cell Size/Type	Fuel Cell Manufacturer	Range (mi/km)	Maximum Speed	Fuel Type
	THINK FC5	Fuel cell	85 kW/PEM	Ballard Mark 900 series	N/A	80 mph/128 km/h	Methanol
	Advanced Focus FCV	Fuel-cell/battery hybrid	85 kW/PEM	Ballard Mark 900 series	180 mi/290 km	N/A	8.8 lb compressed H2 at 5000 psi
	Explorer	Fuel-cell/battery hybrid	60 kW/PEM	Ballard	350 mi/563 km	N/A	10 kg hydrogen at 700 bar
	Airstream concept vehicle/Hyseries Edge	Fuel-cell/plug-in hybrid	HySeries drive	Ballard	280 miles (using hydrogen)—305 total	85 mph	
GM	Sintra (minivan)	Fuel cell	50 kW/PEM	N/A	N/A	N/A	N/A
Hydrogenics works with GM on FC development	Zafira (minivan)	Fuel cell	50 kW/PEM	Ballard	300 mi/483 km	75 mph/120 km/h	Methanol
	Precept FCEV concept only	Fuel-cell/battery hybrid	100 kW/PEM	GM	500 mi/800 km (est.)	120 mph/193 km/h	Hydrogen (stored in metal hydride)
	HydroGen3 (Zafira van)	Fuel cell	94 kW/PEM	GM	250 mi/400 km	100 mph/160 km/h	Liquid hydrogen
	Chevy S-10 (pickup truck)	Fuel-cell/battery hybrid	25 kW/PEM	GM	240 mi/386 km	70 mph	Low sulfur clean gasoline (CHF)

Model	Type	Power/Membrane	Supplier	Range	Speed	Storage
Autonomy concept only	Fuel cell	N/a	N/A	N/A	N/A	N/A
Hy-Wire proof of concept	Fuel cell	94 kW/PEM	GM	80 mi/129 km	97 mph/160 km/h	4.4 lb compressed H2 at 5000 psi
Advanced HydroGen3 (Zafira van)	Fuel cell	94 kW/PEM	GM	170 mi/270 km	~100 mph/160 km/h	6.8 lb compressed H2 at 10,000 psi
Diesel hybrid electric military truck	Fuel-cell APU	5 kW/PEM	Hydrogenics	N/A	N/A	Low-pressure metal hydrides
Sequel	Fuel-cell/battery hybrid	73 kW/PEM	GM	300 mi	N/A	8 kg compressed H2 at 10,000 psi
Equinox	Fuel-cell/battery hybrid	93 kW/PEM		200 mi/320 km	100 mph/160 km/h	
HydroGen4	Fuel-cell/battery hybrid	93 kW/PEM		200 mi/320 km	100 mph/160 km/h	
Provoq	Fuel-cell/battery hybrid			300 mi	100 mph/160 km/h	Compressed hydrogen at 10,000 psi
GM (Shanghai) PATAC — Phonix (minivan)	Fuel-cell/battery hybrid	25 kW/PEM	Shanghai GM	125 mi/200 km/h	70 mph/113 km/h	Compressed hydrogen
Honda — FCX-V1	Fuel-cell/battery hybrid	60 kW/PEM	Ballard Mark 700 series	110 mi/177 km	78 mph/130 km/h	Hydrogen (stored in metal hydride)
FCX-V2	Fuel cell	60 kW/PEM	Honda	N/A	78 mph/130 km/h	Methanol
FCX-V3	Fuel-cell/Honda ultracapacitors	62 kW/PEM	Ballard Mark 700 series	108 mi/130 km	78 mph/130 km/h	26 gal of compressed hydrogen at 3600 psi

—continued

TABLE 11.1 (continued)
Fuel Cell Manufacturers and Vehicle Data

Automaker	Vehicle Type	Engine Type	Fuel Cell Size/Type	Fuel Cell Manufacturer	Range (mi/km)	Maximum Speed	Fuel Type
	FCX-V4	Fuel-cell/Honda ultracapacitors	85 kW/PEM	Ballard Mark 900 series	185 mi/300 km	84 mph/140 km/h	130 L compressed H_2 at 5000 psi
	FCX	Fuel-cell/Honda ultracapacitors	85 kW/PEM	Ballard Mark 900 series	220 mi/355 km	93 mph/150 km/h	156.6 L compressed hydrogen at 5000 psi
	Kiwami concept	Fuel cell	N/A	N/A	N/A	N/A	Hydrogen
	FCX concept vehicle	Fuel cell	100 kW/PEM	Honda	270 mi/434 km	100 mph/160 km/h	Compressed hydrogen
	PUYO concept vehicle	Fuel cell	N/A	N/A	N/A	N/A	N/A
	FCX Clarity	Fuel cell	100 kW/PEM	Honda	354 mi/570 km	100 mph/160 km/h	Compressed hydrogen
Hyundai	Santa Fe SUV	Ambient-pressure fuel cell	75 kW/PEM	UTC fuel cells	100 mi/160 km	77 mph/124 km/h	Compressed hydrogen
	Santa Fe SUV	Ambient-pressure fuel cell	75 kW/PEM	UTC fuel cells	250 mi/400 km	N/A	Compressed hydrogen
	Tucson	Fuel cell	80 kW/PEM	UTC fuel cells	185 mi/300 km	93 mph/150 km/h	Compressed hydrogen
	i-Blue concept	Fuel cell	100 kW/PEM		373 mi/600 km	102 mph/165 km/h	Compressed hydrogen
Kia	Sportage	Fuel cell	80 kW/PEM	UTC fuel cells	185 mi/300 km	150 km/h	Compressed hydrogen

Mazda	Demio (compact passenger car)	Fuel-cell/ultracapacitor hybrid	20 kW/PEM	Mazda	106 mi/170 km	60 mph/100 km/h	Hydrogen (stored in metal hydride)
	Premacy FC-EV	Fuel cell	85 kW/PEM	Ballard Mark 900 series	N/A	77 mph/124 km/h	Methanol
Mitsubishi	SpaceLiner concept only	Fuel-cell/battery hybrid	40 kW/PEM	N/A	N/A	N/A	Methanol
	Grandis FCV (minivan)	Fuel-cell/battery hybrid	68 kW/PEM	Daimler–Chrysler/Ballard	92 mi/150 km	87 mpg/140 km/h	Compressed hydrogen
Nissan	R'nessa (SUV)	Fuel-cell/battery hybrid	10 kW/PEM	Ballard Mark 700 series	N/A	44 mpg/70 km/h	Methanol
Made prototypes w/each fuel-cell stack	Xterra (SUV)	Fuel-cell/battery hybrid	85 kW/PEM	Ballard Mark 900 series and UTC fuel cells	100 mi/160 km	75 mph/120 km/h	Compressed hydrogen
	X-TRAIL (SUV)	Fuel-cell/battery hybrid	75 kW/PEM	UTC fuel cells (ambient-pressure)	N/A	78 mph/130 km/h	Compressed hydrogen at 5000 psi
	Effis (commuter concept)	Fuel-cell/battery hybrid	N/A	N/A	N/A	N/A	N/A
PSA Peugeot–Citroen	Peugeot hydrogen gen	Fuel-cell/battery hybrid	30 kW/PEM	Nuvera	188 mi/300 km	60 mph/100 km/h	Compressed hydrogen
	Peugeot fuel-cell cab "Taxi PAC"	Fuel-cell/battery hybrid	55 kW/PEM	H power	188 mi/300 km	60 mph/100 km/h	80 L compressed hydrogen at 4300 psi

—continued

TABLE 11.1 (continued)
Fuel Cell Manufacturers and Vehicle Data

Automaker	Vehicle Type	Engine Type	Fuel Cell Size/Type	Fuel Cell Manufacturer	Range (mi/km)	Maximum Speed	Fuel Type
	H_2O fire-fighting concept only	Battery/fuel-cell APU	N/A	N/A	N/A	N/A	Catalyzed chemical hydride-sodium Borohydride
Renault	EU FEVER project (Laguna wagon)	Fuel-cell/battery hybrid	30 kW/PEM	Nuvera	250 mi/400 km	75 mph/120 km/h	Liquid hydrogen
Suzuki	Covie concept only	Fuel cell	N/A	GM	N/A	N/A	N/A
	Mobile terrace	Fuel cell	N/A	GM	N/A	N/A	Hydrogen
Toyota	RAV 4 FCEV (SUV)	Fuel-cell/battery hybrid	20 kW/PEM	Toyota	N/A	62 mph/100 km/h	Hydrogen (stored in metal hydride)
	RAV 4 FCEV (SUV)	Fuel-cell/battery hybrid	25 kW/PEM	Toyota	310 mi/500 km	78 mph/130 km/h	Methanol
	FCHV-3 (Kluger V/Highlander SUV)	Fuel-cell/battery hybrid	90 kW/PEM	Toyota	186 mi/300 km	93 mph/150 km/h	Hydrogen (stored in metal hydride)

	Type	Power/Membrane	Manufacturer	Range	Speed	Storage
FCHV-4 (Kluger V/ Highlander SUV)	Fuel-cell/battery hybrid	90 kW/ PEM	Toyota	155 mi/250 km	95 mph/150 km/h	Compressed hydrogen at 3600 psi
FCHV-5 (Kluger V/ Highlander SUV)	Fuel-cell/battery hybrid	90 kW/ PEM	Toyota	N/A	N/A	Low-sulfur clean gasoline (CHF)
FCHV (Kluger V/Highlander SUV)	Fuel-cell/battery hybrid	90 kW/ PEM (122 hp)	Toyota	180 mi/290 km	96 mph/155 km/h	Compressed hydrogen at 5000 psi
FINE-S concept only	Fuel cell	N/A	N/A	N/A	N/A	N/A
VW — EU Capri Project (VW Estate)	Fuel-cell/battery	15 kW/ PEM	Ballard Mark 500 series	N/A	N/A	Methanol
HyMotion	Fuel cell	75 kW/ PEM	Ballard	220 mi/355 km	86 mph/140 km/h	13 gal of liquid hydrogen
HyPower	Fuel-cell/ supercapacitor hybrid	40 kW/ PEM	Paul Scherrer Institute	94 mi/150 km	N/A	Compressed hydrogen
Touran HyMotion	Fuel-cell/battery	80 kW/ PEM		N/A	140 km/h	
Space Up Blue	Fuel-cell/battery	45 kW/ PEM	VW	155 mi/250 km	N/A	

a Created by U.S. Fuel Cells Council (http://www.fuelcells.org/info/charts/carchart.pdf).

Graphite nanofibers are stacked sheets of graphite that can store up to three times their own weight in hydrogen under pressure at room temperature. A 0.34-nm gap exists between the graphite layers. Because hydrogen molecules are 0.26 nm, they fit perfectly into these gaps. At this time, however, much work needs to be done to perfect storage, manufacture, and control of this technology. The chemist's role in this challenge is paramount.

11.6 MEMBRANE ELECTRODE ASSEMBLY

Membrane electrode assemblies (MEAs) are typically five-layer structures, as shown in Figure 10.1. The membrane is located in the center of the assembly and is sandwiched by two catalyst layers. The membrane thickness can be from 25 to 50 μm and, as mentioned in Chapter 10, made of perfluorosulfonic acid (Figure 11.3). The catalyst-coated membranes are platinum on a carbon matrix that is approximately 0.4 mg of platinum per square centimeter; the catalyst layer can be as thick as 25 μm [12]. The carbon/graphite gas diffusion layers are around 300 μm. Opportunities exist for chemists to improve the design of the gas diffusion layer (GDL) as well as the membrane materials. The gas diffusion layer's ability to control its hydrophobic and hydrophilic characteristics is controlled by chemically treating the material. Typically, these GDLs are made by paper processing techniques [12].

The cost imperative comes into play greatly with fuel cell membrane assemblies. It is up to the automotive chemist to design ways to improve the cost structure of these assemblies. For instance, the perfluorosulfonic membrane (Nafion) by DuPont is a very expensive component [13]. Other companies, such as Arkema and PolyFuel, are proposing other films as a cost alternative to the Nafion product [14,15]. The precious metal catalyst used in membrane assemblies is applied to carbon using batch methods; if other deposition methods (such as vapor deposition) can be developed, cost can be reduced [12].

For sealing, there is approximately 1 mile of sealant for an 80-kW fuel-cell stack [12]. Development of a high-speed application will greatly reduce the cost of manufacturing these seals. One avenue of cost reduction could be the use of a thin film manufacturing process, spray coatings, or screen printing [12]. The manufacture of bipolar plates, which eliminates secondary machining operations, is another cost

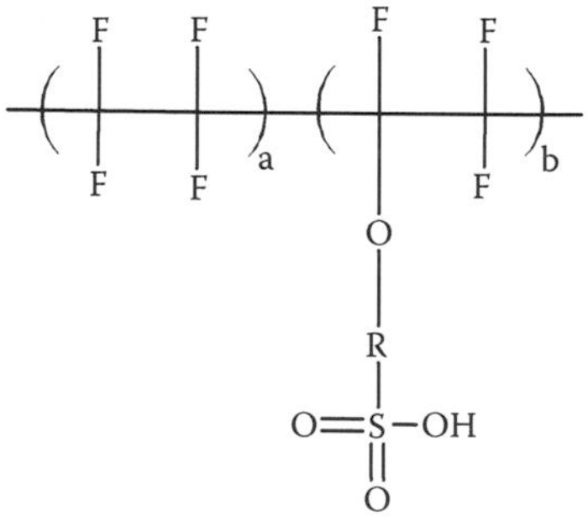

FIGURE 11.3   Structure of perfluorosulfonic membrane.

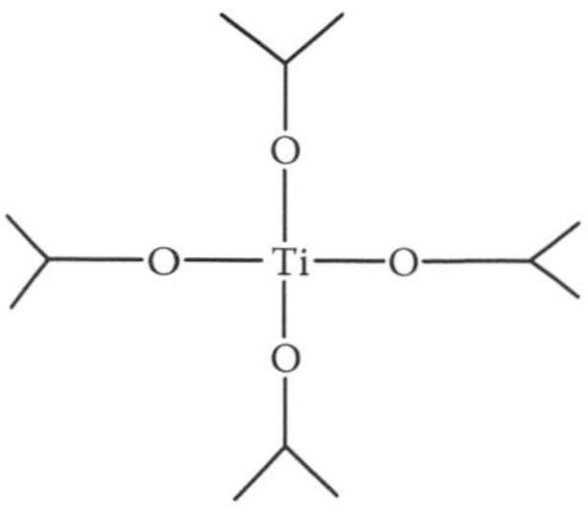

FIGURE 11.4 Structure of titanium isopropoxide.

reduction opportunity. As in some engineering applications, polymer seals such as polyetheretherketones (PEEKs), flatness, and parallelism of the bipolar plate play a role in the performance of the part. The flatness and parallelism affect the uniformity of the flow fields [12]. If tolerances are not controlled in the plates, uneven flow fields can produce uneven distributions of reactants [16]. This will in turn affect the performance and durability of the fuel cell.

Graphite and metal bipolar plates are currently in use by fuel-cell manufacturers, who have proposed rapid, high-rate manufacture of these plates [16]. Injection molding of bipolar plates has been proposed for carbon- and graphite-containing resins [16]. Control and design of the polymer matrix and process parameters will, of course, need to be designed carefully. Metallic bipolar plates will be stamped and controlled through the process.

Ways to increase membrane durability have been examined by various researchers across the country. Maurtiz et al. examined the use of metal-oxide metal particles to increase the properties of the membrane. A titanium isopropoxide (Figure 11.4) addition to Nafion membranes generates quasi-network particles; this improves membrane modulus and dimensional stability [17]. In addition, the titanium matrix reduces fuel crossover and minimizes chemical degradation. Table 11.2 shows the increase in modulus along with stress and strain and stress changes after the addition of the titanium matrix [17]. With a 20% load of the titanium matrix, performance criteria remain comparable. Acid functionality remains intact; however, water uptake is reduced as volume inside clusters is occupied. Conductivity is reduced due to chain mobility [17].

TABLE 11.2

Properties of Nafion and Nafion/Ti Membranes

	Modulus (MPa)	Stress at break	Strain at break
Nafion/Ti isopopoxide	120.4 ± 7.1	24.1 ± 1.6	3.1 ± 2
Nafion	36.2 ± 7.2	20.8 ± 3.2	4.1 ± 4

11.7 CELL STACK ASSEMBLY

As many as 400 membrane electrode assemblies or stacks are applied together to make up a fuel cell. Alignment of the cells will be required to be within ±5 μm of one another [12]. Stress will be lowered by the tighter requirements and the fragile carbon assembly will gain some necessary protection. This requirement will make speed of assembly difficult, which in turn will result in higher cost.

REFERENCES

1. http://www.cem.msu.edu/~cem181h/projects/97/fuelcell/chem1.htm
2. http://www.pharmaceutical-echnology.com/contractors/lab%5Fequip/whitehouse/
3. http://www.emerson.com/en-US/Pages/Home.aspx
4. http://en.wikipedia.org/wiki/Glass_microspheres
5. http://www.inrf.uci.edu/research/posters/Nanoparticle_Synthesis_with_Air-Assisted_Ultrasonic_Spray_Pyrolysis_Prof_Chen_Tsai_UCI_949-824-5144_HJ_Yoo.pdf
6. http://www.casdn.neu.edu/chronicle/sp97-4.html
7. Corio, P., A. P. Santos, P. S. Santos, M. L. A. Temperini, V. W. Brar, M. A. Pimenta, and M. S. Dresselhaus. 2004. Characterization of single wall carbon nanotubes filled with silver and with chromium compounds. *Chemical Physics Letters* 383:475.
8. Zhao, L., and L. Gao. 2004. Filling of multi-walled carbon nanotubes with tin (UV) oxide. *Carbon* 42:3251.
9. Dillion, A. C., K. E. H. Gilbert, P. A. Parilla, J. L. Allerman, G. L. Hornyak, K. M. Jones, and M. J. Heben. 2002. Hydrogen storage in carbon single-wall nanotubes. In *Proceedings of the 2002 U.S. DOE Hydrogen Program Review* NREL/CP-610-32405.
10. Neskovic, O., J. Djustebek, V. Djordjevic, J. Cvticanin, S. Velickovic, M. Veljokovic, and N. Bibic. 2006. Hydrogen storage on activated carbon nanotubes. *Digest Journal of Nanomaterials and Biostructures* 1 (4): 121–127.
11. Gommans, H. H., J. W. Alldredge, H. Tashiro, J. Park, J. Magnuson, and A. G. Rinzler. 2000. Fibers of aligned single-walled carbon nantotubes: Polarized Raman spectroscopy. *Journal of Applied Physics* 88:2509.
12. Roadmap on Manufactuing R&D for the Hydrogen Economy. 2005. Manufacturing research & development of PEM fuel cell systems for transportation applications. Background material for the Manufacturing R&D Workshop, Washington, D.C. Department of Energy. pp. 3, 4. http://www.hydrogen.energy.gov/pdf/roadmap_manufacturing_hydrogen_economy.pdf
13. Carlson, E. 2003. Cost analysis of fuel cell stacks/systems. 2003. Hydrogen and Fuel Cells Merit Review Meeting, Berkley, CA. Precious Metal Availbility and Cost Analysis for PEMFC commercialization. http://www1.eere.energy.gov/hydrogen and fuel cells/pdfs/merit 03/104_tiax_eric_carbon.pdf
14. Gaboury, S. 2004. Development of a low-cost, durable membrane and MEA for stationary and mobile fuel cell applications. Atofina Chemicals, Inc. 2004 Annual Program Review, Philadelphia, PA.
15. http://www.polyfuel.com/technology/hydrogen.html
16. Kelly, K. J. et al. 2005. Application of advanced CAE methods for quality and durability of fuel cell components. 2005 Annual Program Review Meeting. May 23–26 2005 Arlington, VA. http://www.hydrogen.energy.gov/annual_review05_fuelcells.html
17. http://www.hydrogen.energy.gov/pdfs/review08/fcp_11_mauritz.pdf

12 Developing Technology

12.1 INTRODUCTION

The chemist will play an expanded role in future automotive design and technology. This is largely due to the increased role desired from the four imperatives (performance, mass, cost, and environmental effects). As discussed in Chapter 4, the imperatives are interrelated to a certain extent and none is conceptually more important than another. With oil prices rising ($141.37/barrel) with no relief in site, the imperatives of cost and environment are of grave concern [1]. To deal with this, it is necessary for the chemist to play a role in new technologies. The most promising technologies for the immediate future are those concerned with hybrid vehicles, electric vehicles (including battery technology), ozone reduction, and biomaterials.

12.2 HYBRID TECHNOLOGIES

Hybrid technologies within OEMs (original equipment manufacturers) are considered to be a combination of two or more sources of power. These can include batteries (or rechargeable energy storage systems), fuel cells, internal combustion engines, and electric motors. For example, a moped that can utilize the mechanical energy of the rider peddling the bicycle along with a small internal combustion engine mounted on the frame utilizes two different power sources.

Diesel–electric railway locomotives are also hybrid technologies because the diesel engine will drive an electric generator that will power an electric motor for locomotion. Bombardier has developed the AGC (*Autoraila Grande Capacite*), which is a dual-mode diesel electric as well as a dual-voltage railcar. The system can operate on 1,500–2,500 V [2].

The Kiha E200 in Japan utilizes lithium ion batteries that are mounted within the vehicle [3]. The significance of the design is that this is the first operational rail system to utilize energy storage and regeneration. General Electric is also working on what is called the "Ecomagination"—a system that stores energy using $NaNiCl_2$ batteries rather than lithium ion batteries. In addition, these $NaNiCl_2$ batteries capture the energy from dynamic braking and coasting down hills. General Electric is investing heavily (up to around $2 billion) in the research and expects a 10% fuel savings reduction [4].

Other variants exist for rail hybrid technology, such as Canada's Railpower technologies GG and GK. These technologies utilize long-life, rechargeable lead acid batteries to power 1,000- to 2,000-hp motors. Obviously, the imperative of environment is affected due to the decrease in diesel emissions that is possible from this type of system (up to an estimated 40–60% diesel savings) [5].

An economic crossroads or favorable business case was reached in 2008 with fuel costs going over the $140-per-barrel mark as far as commercial automotive vehicles were concerned. General Motors, for instance, launched several hybrid versions of its full-sized truck. Until that point, an unfavorable business case had existed for introducing hybrids into the mainstream commercial vehicle market—that is, the extra cost associated with manufacturing a hybrid system along with its extra content did not offset the fuel savings. In addition, battery technology cost improvements coupled with the increase in gasoline prices and the increased customer concerns about the environment drove their development.

Toyota, GM, Ford, and Kenworth have all introduced hybrids. For instance, the Kenworth T270 Class 6 is an example of a large (medium-duty-plus) application that is competitive with traditional vehicles [6]. Some other hybrids are GM's Chevy Tahoe, GMC Yukon, Chevrolet Silverado, Cadillac Escalade, Saturn Vue, Toyota Prius, Highlander, Camry hybrid, Ford Escape, Honda Insight, and Honda Civic hybrid. Some performance data are listed in Table 12.1 [7]. The greater mileage values from city driving are one of the major benefits of hybrid technology. These benefits are established by several factors including:

- shutting down the internal combustion engine (ICE) during traffic stops, coasting, or idling;
- recapturing energy from braking (i.e., regenerative braking);
- improved aerodynamic effects;
- relying on the ICE as well as the electric or hybrid motor for peak power periods;
- utilizing lower rolling resistance tires (hybrids utilize a more inflated tire that reduces resistance by 50%);
- having greater storage capacity and reuse of recaptured energy; and
- utilizing RESS (rechargeable energy storage), a system that stores the energy for delivery of electric energy (i.e., electric batteries or ultracapacitors).

Figure 12.1 shows a schematic of a plug-in style electric hybrid system. It uses an off-board system to recharge the batteries and has a generator powered by an internal

TABLE 12.1
Hybrid Vehicle Performance

Hybrid Model	Mileage (mpg)[a]	Horse Power
2008 Honda Civic hybrid	47 city/48 hwy	110 hp 1.3 L SOHC 4 cylinder
2007 Honda Accord hybrid	29 city/37 hwy	240 hp 3.0 L V-6
2008 Nissan Altima hybrid	42 city/36 hwy	158 hp 2.5 L 4 cylinder
2008 Toyota Prius	46 combined city/hwy	110 hp gas and electric combined
2008 Ford Escape hybrid	34 city/30 hwy	155 hp gas and electric combined

[a] Data from Eartheasy (http://www.eartheasy.com/live_hybrid_cars.htm).

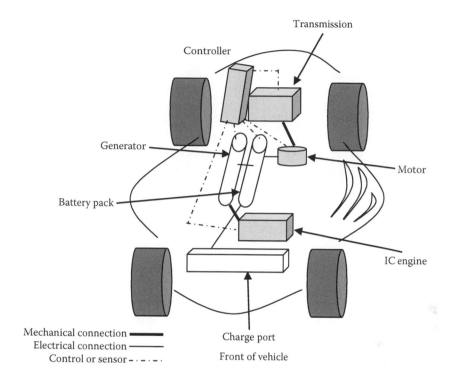

FIGURE 12.1 Diagram of a plug-in hybrid electric vehicle (PHEV).

combustion engine for range extension. This type of system is a series hybrid. In a parallel hybrid, the engine will power the vehicle with some assistance from the electric motor.

With plug-in hybrid electrical vehicles (PHEVs), the gasoline–electric hybrid will have a battery pack (lithium ion) that has a large capacity and can be recharged. Some of these latest designs run on battery power for up to 60 miles. When the battery is practically discharged, the vehicle will switch to the gasoline engine and recharge the battery. The advantage of PHEV vehicles is that recharging stations and thus battery range issues can be avoided by utilizing this type of hybrid.

Some issues associated with this type of system include finding an inexpensive battery pack (discussed later in this chapter) as well as concurrent charging of systems. For instance, if an entire sector charges vehicles by simultaneous connection into the energy grid, then the coal-generated electrical source would be a problem. The U.S. Department of Energy, as well as OEMs, continues to investigate the best ways to implement PHEVs [9].

In 2007 at the Detroit Auto Show, General Motors Corporation introduced a PHEV that utilizes a lithium ion battery [9]. The Volt is a plug-in series hybrid that is expected to be produced in 2010 [10]. The average distance the American commuter travels today is approximately 64 km, or 40 miles, according to *US News and World Report* [11]. The Volt design is set to run purely on electrical power for the first 40 miles, which will cover the range of the average commute in the United States. The range of these lithium ion vehicles is about 1,000 km on the highway with use of the ICE. Production of these vehicles could reach 60,000 units by 2011 [12].

General Motors also introduced the E-flex drive system, which makes many of the components in an electric hybrid vehicle common so that the systems are interchangeable. The voltage output of the lithium ion batteries can be adjusted to meet requirements by adjusting the number of cells. The primary engine on some of these PHEV vehicles is typically a 1.0-L, turbocharged, three-cylinder engine with flexible fuel capability. The flexible fuel is typically a mixture of 85% ethanol and 15% gasoline. Charging from a North American 120-V, 15-A outlet would take approximately 6.5 hours.

Other options to help the environment imperative could be utilized for these types of hybrids. This includes using a pure ethanol (E100) ICE, a fuel cell, or a biodiesel fuel engine.

12.3 BIODIESEL

A biodiesel or alkyl ester fuel engine can be used within a hybrid system or alone. Biodiesel is a clean-burning alternative to diesel or petroleum. It is produced from fats or oils through transesterification. Biodiesel must be produced through ASTM D6751 in order to ensure its performance [13]. Biodiesel produced in this manner can be used with little modification to its structure. Sulfur or aromatics are not present in its structure. The transesterification process used to produce the fuel is shown in Figure 12.2.

The biodiesel production process has three basic routes from fats and oils to produce esters or biodiesel, according to the National BioDiesel Board [14]:

- base catalyzed transesterification of the oil with alcohol;
- direct acid catalyzed esterification of the oil with methanol; and
- conversion of the oil to fatty acid catalysis.

In the United States, the most utilized process is the base catalyzed transesterification of oil with alcohol. The base catalysis is popular because of its

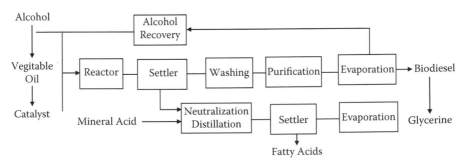

FIGURE 12.2 Biodiesel production process.

economy—no intermediates are produced in the reaction and conversion to methylester is direct. The reaction takes place at 150°F at a pressure of 20 psi [14]. A fat or oil is reacted with an alcohol in the presence of a catalyst to produce glycerin and methylester or biodiesel [13]. Typically, the catalyst is NaOH or KOH in a methanol solution [13]. Product and reactant amounts are shown in Table 12.2. The products produced from the reaction show that 100% of the yield is utilized. The efficiency of this process is what makes it the best choice for the manufacture of biodiesel.

Production of biodiesel fuel in the United States since 1999 is shown in Table 12.3 [13]; it is expected to grow in the coming years. Biodiesel has completed the health effects testing requirements of the 1990 Clean Air Act and is legally registered with the Environmental Protection Agency as a legal motor fuel for sale and distribution within the United States.

Biodiesel is technically defined as monoalkyl esters of long chain fatty acids derived from vegetable oils or animal fats conforming to ASTM D6751 specifications [13]. Blends are denoted as "BXX"; XX indicates the percentage of biodiesel contained within the blend. As we go forward, it is up to today's automotive chemist to work to develop efficient processes such as these in order to maintain our environment as well as to remain competitive in a global market.

TABLE 12.2
Products and Reactants in Biodiesel

Reactant (%)	Product (%)
Alcohol (12)	Alcohol (4)
Catalyst (1)	Glycerin (1)
Oil (87)	Fertilizer (9)
	Methylester (86)

TABLE 12.3
Biodiesel Production in the United States

Quantity (gal)	Year
250 Million	2006
75 Million	2005
25 Million	2004
20 Million	2003
15 Million	2002
5 Million	2001
2 Million	2000
0.5 Million	1999

12.4   BATTERY TECHNOLOGIES

Much of the success of the new hybrid, flex fuel, and electric vehicle technologies depends on the chemical knowledge of current and future engineers and chemists. This is shown best in the successful implementation of battery technology. General Motors executives have stated that battery technology will have a large role in determining the impact of hybrid and electric vehicles [15]. Lithium ion technology will be utilized on the GM's Volt concept. Suppliers to the OEM are Compact Power, which uses manganese oxide-based cells made by LG Chemical, and Continental Automotive, which uses nanophosphate cells made by A123 Systems [16,17]. Expected life of the lithium ion batteries is 10 years.

12.5   LITHIUM ION BATTERY

With lithium ion battery (or Li-ion) technology, a lithium ion moves between the cathode and the anode in a rechargeable system. When the system is charging, the lithium ion moves from the cathode to the anode, and when the system discharges, the lithium ion moves from the anode to the cathode. Lithium ion batteries have excellent energy-to-weight ratios, along with a slow loss of charge when they are not in use; these make them good candidates for automotive applications.

Various materials are used for production of the three main components of a lithium ion battery. Research and development of these materials is where the automotive chemist is severely needed. The main components of the battery are the electrolyte, cathode, and anode. For the cost imperative, graphite is used most often in the anode. The cathode is typically a layered lithium cobalt oxide, lithium iron phosphate, or lithium manganese oxide. Other materials, such as TiS_2, have been used [18]. Of course, properties vary depending on the choice of anode, cathode, electrolyte, etc.

Lithium batteries were first made in the 1970s by the Exxon Corporation [19]. These first batteries contained metallic lithium used at the anode, which presented safety issues. Lithium ion batteries are considered to be secondary batteries (rechargeable) that contain different layers of anode material. It was not until the development of the graphite anode that lithium ions became feasible. The cathode material that was utilized in these batteries was spinel, which is a mineral with the formula XY_2O_4. The class of minerals was named after the magnesium variant or spinel ($MgAl_2O_4$). This family of compounds also includes gahnite ($ZnAl_2O_4$), franklinite (($Fe,Mn)_2O_4$), chromite (($Fe,Mg)Cr_2O_4$), and magnetite (Fe_3O_4). They crystallize in a cubic close-packed lattice crystal system; the cations occupy the octahedral and tetrahedral sites in the lattice. Because the spinel material will lose some properties over time, battery chemists have made some modifications to alleviate losses.

For lithium ion batteries, the half reactions are

Anode:
$$xLi^+ + xe^- + 6C \leftrightarrow Li_xC_6$$

Cathode:
$$LiCoO_2 \leftrightarrow Li_{1-x}CoO_2 + xLi^+ + xe^-$$

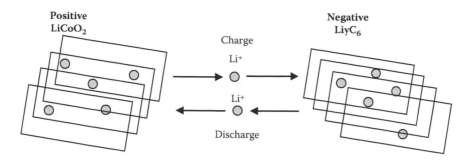

FIGURE 12.3 Lithium ion flow in lithium ion battery system.

The general term for the process of moving from anode to cathode by the lithium ion or insertion is *intercalation*. The reverse of this process—moving from cathode to anode—is called extraction or deintercalation. Figure 12.3 shows how lithium ions are transported to and from the cathode and anode. The cobalt in the positive electrode is being oxidized from the Co^{+3} state to the Co^{+4} state during the charge phase and reduced from Co^{+4} to Co^{+3} during the discharge phase. The open crystal structure allows extraction of the lithium ions. As stated, the cathode has many possible chemistries, including manganese dioxide. Table 12.4 shows the average voltages of some possible cathode materials.

When discharge occurs, lithium ions from the anode migrate across the electrolyte to the cathode. Migration of electrons across the circuit will balance the reaction. The migration occurs in an organic solvent such as ether because these cells contain no aqueous media. Lithium produces a violent reaction with water. The electrolyte is lithium salts, typically $LiPF_6$ or $LiClO_4$. Unfortunately, decomposition of the solvent occurs on the anode during charging. When decomposition does occur, a solid electrolyte interphase (SEI) forms [20]. This SEI prevents decomposition of the electrolyte after the second charge.

The mass of lithium ion batteries allows the automotive chemist and design engineers to decrease the mass of the overall vehicle as well as increase the performance of the vehicle. They are much lighter than other rechargeable battery designs. Lithium ions are also a wise choice for automotive use because self-discharge is very low (5% per month vs. 30% for nickel–metal hydride [NiMH] batteries).

OEM research around lithium ion technology is ongoing and many issues still need to be overcome. Lithium ion batteries have a life span that is age dependent,

TABLE 12.4
Average Voltage for Cathode Materials

Material	Average Voltage (V)
$LiCoO_2$	3.7
$LiMnO_2$	4.0
$LiFePO_4$	3.3
Li_2FePO_4F	3.6

TABLE 12.5
Lithium Ion Loss versus Storage Temperature

Storage Temp. (°C)	40% Charge (% Loss after 1 Year)	100% Charge (% Loss after 1 year)
0	2	6
25	4	20
40	15	35
60	25	40

unlike many other battery systems. The storage temperatures are highly interactive with the life span of lithium ion batteries. In addition, the state of charge as well as the temperature can determine the capacity loss of the battery. For instance, a battery stored at 40% charge will last longer than one stored at 100% charge at a higher temperature. The permanent loss of capacity at various temperatures at different states of charge is shown in Table 12.5 [21].

As can be seen from the table, battery storage at 40% versus storage at full charge is preferable. For an automotive application, storage will occur at 0–25°C most of the time. For a vehicle such as the Volt, the state of desired charge if the system is being regenerated by an ICE is unclear. Most likely, a circuit will exist (as does now on all current Li-ion batteries) that will shut down the system if it falls below a certain state of charge. Other issues include the inherent danger over nickel–metal hydride designs, which forces safety measures within the battery to protect against extreme pressure, heat, and overcharging. Li-ions can explode if temperatures are too high.

12.6   NICKEL–METAL HYDRIDE CELLS

In nickel–metal hydride (NiMH) cells, a hydrogen-absorbing alloy is used for the anode. The chemistry is dependent on the ability of the anode to absorb a large quantity of hydrogen. Hydrides store hydrogen, which reacts reversibly in the battery cell. Nickel oxyhydroxide (NiOOH) is used for the cathode. The electrolytes used in an aqueous solution are compounds such as potassium hydroxide. These types of systems are similar to nickel–cadmium, but the hydrogen absorbs alloy for the anode; this gives two to three times the storage capacity of that of a similar nickel–cadmium battery. These batteries are safer than lithium ion batteries, but the energy density is lower, which makes them undesirable from a performance point of view.

Development of these batteries took place at around the same time frame as lithium ion technologies in the 1980s. The NiMH type battery was used in the General Motors EV1, as well as other plug-in vehicles. The Toyota RAV4 EV, Honda EV Plus, Ford Ranger EV, and Honda Insight all use NiMH batteries. The reactions for nickel–metal hydrides are

Anode rxn: $H2O + M + e^- \rightleftharpoons OH{-} + MH$

Cathode rxn: $Ni(OH)2 + OH^- \rightleftharpoons NiO(OH) + H2O + e{-}$

In the anode reaction, charge takes place from left to right and discharge occurs in the opposite direction. The metal is an intermetallic AB_5-type compound. Some variation occurs in that a rare-earth mixture (typically cerium or lanthanum) is compounded with nickel, manganese, cobalt, or aluminum (for the B portion). Titanium and vanadium can also be used in AB_2 intermetallic compounds with nickel, zirconium cobalt, or chromium (as the B portion), but they are rarely used due to performance issues.

The reversible metal hydrides produce oxygen at the positive electrode when overcharged at a reasonably low rate. The oxygen will go through a separator and recombine at the surface of the anode. With hydrogen suppression at the anode, charging energy is converted to heat, thereby allowing the battery to remain sealed.

Batteries are charged at a rate of 1 C (C is the capacity of the battery in amperehours). One is the multiplier in the equation used for calculating charging current (amps = C * multiplier). After 100% charge, the battery begins to overcharge; voltage polarity will reverse and a decrease in voltage will occur. The battery charger will sense the voltage drop and end the charge cycle. Of course, time and temperature affect the cycle.

Recent changes to NiMH batteries have improved their historically weak internal leakage issues. The internal leakage (self-discharge) rate stabilizes around 0.5–1% per day at room temperature [23]. The "low self-discharge batteries" for small applications claim to retain up to 85% of their capacity after a year when stored at room temperature. However, for auto applications, discharge rates along these lines are not yet known.

Nickel–metal hydrides have lower energy density than lithium ion batteries [24]. Cobasys is a major manufacturer of nickel–metal hydride batteries. According to Boschert, Cobasys has large-format NiMH batteries but has not manufactured or licensed the technology for automotive purposes [25]. This was apparent when Toyota employees had difficulty acquiring smaller orders for service for the existing RAV-4EVs. However, GM and Cobasys stated that the Saturn Aura hybrid would use an NiMH battery [26].

12.7   BATTERY DEVELOPMENTS

To extend the range and improve the safety of lithium ion batteries, Valence Technology in Austin, Texas, developed iron phosphate cathodes ($LiFePO_4$) [27], which are produced under license from MIT. These cathodes allow for operation at temperatures up to 60°C. They are now in use in PHEV applications. The researchers at MIT utilized a process that produces nanosize wires, which are used to manufacture thin lithium ion batteries that will have increased energy density [28]. In 2007, fluorine was used to replace the hydroxide group in the iron phosphate cathode by researchers at the University of Waterloo in Canada [29]. The technology promises less volume change and thus greater battery life, as well as increased energy density. Ongoing research by automotive chemists has mostly been directed toward the limiting factors of cathode materials.

Battery experts recommend tailoring a battery's design to fit the application; otherwise, a battery may be designed for small size and long run time but will have a limited life cycle [30]. Buchmann summarized the strengths and limitations of popular batteries and the results are shown in Table 12.6 [29]:

- *Nickel–cadmium* batteries have moderate energy density. These batteries are used for long life and higher discharge rates; extended temperature ranges are required.
- *Nickel–metal hydride* batteries have higher energy densities as compared to nickel–cadmium batteries, but at the expense of reduced cycle life [29].
- *Lead acid* batteries are the most inexpensive option for larger powered applications. They are utilized when mass is not as important, such as in wheelchair applications.
- *Lithium ion* technology offers high-energy density with the added benefit of low mass. These characteristics make it ideal for automotive applications. This technology is the fastest growing battery system [29]. Protection circuits are needed to limit voltages for safety reasons.

12.8   DIRECT OZONE REDUCTION SYSTEMS

In 2002, California Assembly Bill 1493 was signed into law. This bill directed the California Air Resources Board (CARB) to develop new regulations to reduce greenhouse gas (GHG) emissions starting in 2009. In September of 2004, the regulations were developed with an effective date of January 1, 2006. Table 12.7 shows California emission standards by durability level and emission amount. In 1998, CARB adopted a general rule allowing NMOG (nonmethane organic gas) credit to a system developed by the BASF Corporation for ozone reduction. The system is basically an ozone-to-oxygen catalyst applied to one of the external heat exchangers (radiator) of a vehicle. California legislation requires 30% of fleet vehicles to be at a PZEV (partial zero emissions vehicle) level of evaporative and tail pipe emissions. For PZEVs and SLEVs (superlow emissions vehicles), that level is 10 mg/mile of nonmethane organic gas. It is up to the automotive chemist to work with HVAC suppliers, coating suppliers, raw material manufacturers, etc., on forward thinking ideas such as this to achieve all environmental goals set forth by OEMs and governmental authorities.

As discussed in Chapter 9, oxygen has three allotropes or forms in the ozone–oxygen cycle: atomic oxygen (O), ozone (O_3), and diatomic oxygen (O_2). Ozone is formed when oxygen photodissociates after absorbing ultraviolet radiation of wavelengths < 240 nm. The oxygen atoms produced then combine with atomic oxygen to produce ozone. Several free radical catalysts, such as the hydroxyl radical, nitric oxide, and halogens, will convert ozone; however, the BASF catalyst is proprietary. Key to the chemist's and engineer's tasks is to make certain that any environmental improvements made by coatings do not hurt the performance of the heat exchanger. Often a delicate balance must be obtained between the imperatives of performance and environmental considerations.

TABLE 12.6
Properties of Some Batteries

	Nickel–cadmium	Nickel–metal Hydride	Lead Acid (Sealed)	Lithium Ion (Cobalt)	Lithium Ion (Manganese)	Lithium Ion (Phosphate)
Gravimetric energy density (Wh/kg)	45–60	60–120	30–50	150–190	100–135	90–120
Internal resistance (mW)	100–200 (6 V)	200–300 (6 V)	<100 (12 V)	150–130 (cell)	25–75 (cell)	25–50 (cell)
Cycle life (to 80% of capacity)	1500	300–500	200–300	300–500	>500	>1000
Fast charge time	1 h	2–4 h	8–16 h	1.5–3 h	<1 h	<1 h
Overcharge tolerance	Moderate	Low	High	Low; no trickle charge	Low; no trickle charge	Low; no trickle charge
Self-discharge/month at room temp. (%)	20	30	5	<10	<10	<10
Cell voltage (nominal average; V)	1.25	1.25	2	3.6	3.6	3.3
Load current (peak; best result)	20 C: 1 C	5 C: 0.5 C	5 C: 0.2 C	<3 C: 1 C	>30 C: 10 C	>30 C: 10 C
Operating temp. (°C)	−40 to 60	−20 to 60	−20 to 60	−20 to 60	−20 to 60	−20 to 60
Maintenance	30–60 days	60–90 days	3–6 months	Not required	Not required	Not required
Safety	Thermally stable; fuse recommended	Thermally stable; fuse recommended	Thermally stable	Mandatory protection circuit; stable to 150°C	Mandatory protection circuit, stable to 250°C	Mandatory protection circuit; stable to 250°C
Commercial use since:	1950	1990	1970	1991	1996	2006
Toxicity	Highly toxic; harmful to environment	Low toxicity; recyclable	Toxic lead and acids; harmful to environment	Low toxicity; can be disposed in small quantities	Low toxicity; can be disposed in small quantities	Low toxicity; can be disposed in small quantities

TABLE 12.7
California Emission Standards

Category	Durability (Miles)	NMOG	CO	NO$_x$	PM
TLEV	50,000	0.125	3.40	0.40	
	120,000	0.156	4.20	0.60	0.04
LEV	50,000	0.075	3.40	0.50	
	120,000	0.09	4.20	0.70	0.01
ULEV	50,000	0.04	1.70	0.50	
	120,000	0.055	2.10	0.70	0.01
SLEV	120,000	0.01	1.00	0.20	0.01
SLEV+	150,000	0.01	1.00	0.20	0.01

The California requirements state that a vehicle must have onboard diagnostics (OBD) capabilities to receive credit for an ozone coating system. OBD is basically the system within a vehicle's computer that monitors such things as engine performance, O$_2$ sensors, and throttle position sensors. This requirement has been effective since 2006. The level of sophistication can be functionally or performance based. If the OBD system is performance based, it will measure the effectiveness of the actual coating by means of a sensor mounted on the radiator that will communicate to the engine's computer. If the system is a functionally based one, then a simple check indicating some level of ozone reduction is present (i.e., whether the radiator is equipped with the coating). If credit given is less than half of the super-ultralow emissions vehicle (SULEV) requirements (<5 mg/mile), then only a function check is required.

This type of coating can be used to offset evaporative or tailpipe emissions. OEMs such as Audi, Volvo, Mitsubishi, and BMW have already implemented such a system. Some of their models may be found on the CARB Web site [30]. The south coast air basin (SCAB), which utilized the urban air shed model (UAM), was used to model the benefits of such a system. A standard or midsized heat exchanger that would be equivalent to 0.29 m^2 area, 0.4% airflow to vehicle speed ration, and 80% O$_3$ conversion was used. This model was equivalent to 9 mg/mile of NMOG emissions reduction. For SLEV requirements, a 150,000 mile bogey with zero evaporative emissions must be met. Tailpipe emissions should be <0.01 g/mile and total evaporative emissions should be <0.35 g/test. A valuation credit calculation is based on the following equation:

$$Vehicle\_Credit = \left(base\_case\_vehicle\_credit\right) * \left(ratio\_of\_vehicle\_airflow\_to\_base\_case\right) *$$

$$\left(ratio\_of\_vehicle\_conversion\_to\_base\_case\right) * \left(\frac{ratio\_of\_vehicle\_radiator\_front}{surface\_area\_to\_base\_case}\right) *$$

$$\left(ozone\_conversion\_deactivation\_factor\right)$$

$$(12.1)$$

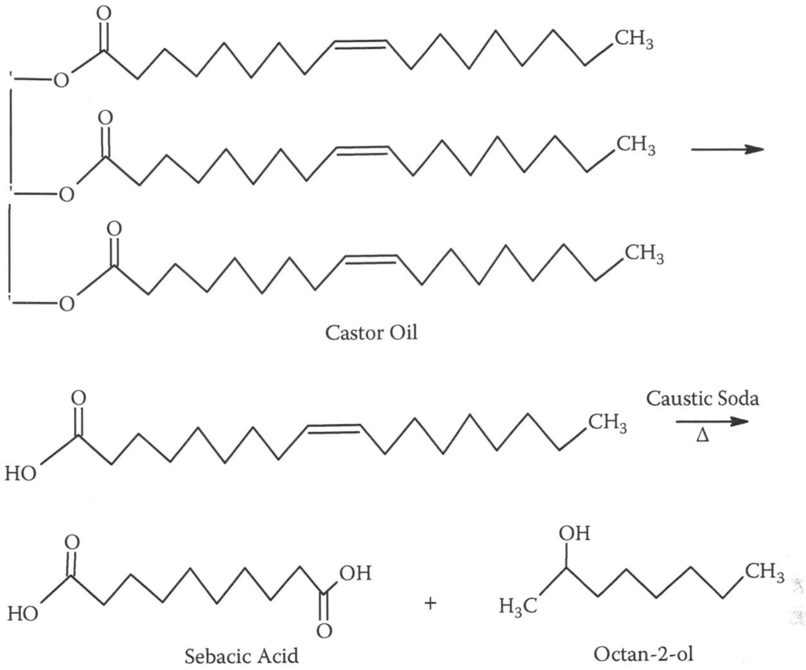

Castor Oil

Caustic Soda
Δ

Sebacic Acid   +   Octan-2-ol

With the base case considered, the equation for credit is

$$= 0.009 \text{ g/mile} * (y/0.4) * (y/80\%) * (z/0.29 \text{ m}^2) * DF \tag{12.2}$$

Equation 12.2 reflects a base case of 9 mg/mile with a 0.29 m² surface area, ratio of airflow to vehicle speed at 40%, and conversion of base case to 80% from the UAM. The deactivation factor (DF) is DF = (aged ozone conversion/ fresh ozone conversion). To determine the deactivation factor, on-road aging of radiators coated with catalyst was run for 150,000 miles and DFs at that point were determined. Freshly coated radiators were exposed to an accelerated aging test and DFs were calculated and compared for accuracy. Figure 12.4 shows the calculation of deactivation factors for two different radiators with different core geometries.

The official credit for a specific OEM vehicle radiator is calculated by comparing the ozone conversion radiator air flow and radiator surface area to that of the base case model utilized by the UAM. The MAC (manufacturer's advisory correspondence) 99-06 certifies the procedure of direct ozone reduction technologies. Of course, the chemist, engineer, and automotive team need to work together to develop the business plan for an effective cost solution to implement such strategies.

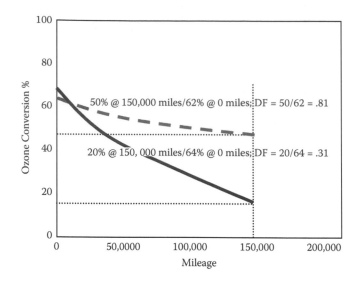

FIGURE 12.4   Deactivation factors calculation chart.

12.9 BIOMATERIALS

Another opportunity for the automotive chemist to improve the environmental imperative is to utilize bio-based materials as building blocks. An outstanding example is the manufacture of nylon 6-10 from castor oil. The BASF Corporation has plans to utilize the product as an alternative to nylon 6. The castor oil is treated with caustic soda and heat to render the sebacic acid portion of the polymer (Equation 12.2).

The sebacic acid is then polymerized with hexamethlyene diamine as with traditional nylons. Figure 12.5 shows the structure of this polymer. Up to 63% of the material can be derived from castor oil. Table 12.8 shows some important properties as compared to nylon 6. The main properties of tensile strength and modulus, at least in a dry state, are fairly similar. Again, we see how the automotive chemist's input can be utilized in vehicle design and development.

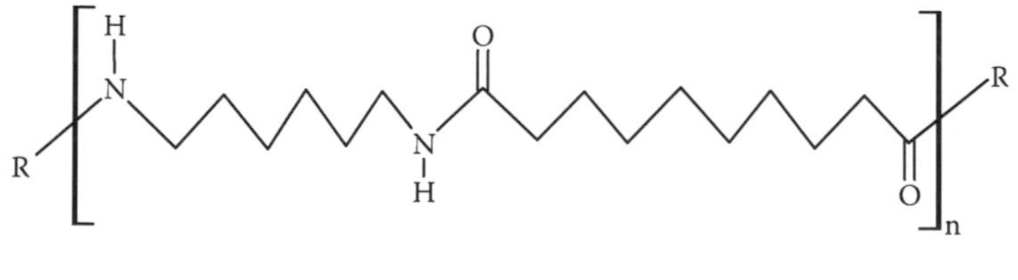

FIGURE 12.5   Structure of nylon 6-10.

TABLE 12.8
Properties of Nylon 6-10 and Nylon 6

	Polyamide 6-10 (MPa)	Nylon 6 (MPa)
Tensile strength	60	80
Modulus of elasticity	2400	3000

Note: Dry properties.

REFERENCES

1. http://www.wtrg.com/daily/crudeoilprice.html (data retrieved July 7, 2008).
2. Needham, J., ed. 1965. *Science and civilization in China,* vol. 4, part 2, 276. Cambridge, England: Cambridge University Press.
3. Japan to launch first hybrid trains. October 29. 2007. *Sydney Morning Herald.*
4. Shabna, J. October 25, 2007. GE's hybrid locomotive: Around the world on brakes. *Ecotality Life.* ecotality.com
5. RailPower Technologies Corp. July 12, 2006. GG series: Hybrid yard switcher. railpower.com
6. Thomas, J. March 27, 2007. Hybrid truck unveiled by Kenworth. treehugger. com
7. http://www.eartheasy.com/live_hybrid_cars.htm
8. U.S. Department of Energy. Office of FreedomCAR and Vehicle Technologies Energy Efficiency and Renewable Energy. 2006. Summary report: Discussion meeting on plug-in hybrid electric vehicles.
9. Eisenstein, P. 2007. GM plugs fuel cells into Volt. thecarconnection.com
10. http://gm-volt.com/2008/06/05/moving-the-chevy-volt-to-production-status/
11. *US News and World Report.* September 24, 2005. Money: New benefit: Help with commuting cost.
12. EV World Newswire. 2007. GM could build 60,000 Volt electric cars in first year. evworld. com
13. http://www.biodiesel.org/resources/faqs/
14. http://www.biodiesel.org/pdf_files/fuelfactsheets/Production.PDF
15. *The Wall Street Journal.* January 11, 2008. Race to make electric cars stalled by battery problems. Online.WSJ.com
16. General Motors. 2007. GM awards advanced development battery contracts for Chevy Volt e-flex system. Press release
17. GM testing Volt's battery, iPhone-like dash on track to 2010. April 4, 2008. *Popular Mechanics.* popularmechanics.com.
18. Thackeray, M. M., J. O. Thomas, and M. S. Whittingham. 2002. Theme article—Science and applications of mixed conductors for lithium batteries.mrs.org
19. Whittingham, M. S. 1976. Electrical energy storage and intercalation chemistry. *Science* 192(4244):1126–1127.
20. Balbuena, P. B., and Y. X. Wang, eds. 2007. *Lithium ion batteries: Solid electrolyte interphase.* London: Imperial College Press.
21. Buchmann, I. 2006. Will lithium-ion batteries power the new millennium? Cadex Electronics Inc.
22. What's the best battery? November 2006. Batteryuniversity.com

23. Review: Testing Sanyo's Eneloop Low Self Discharge chargeable battery. stefanv.com/electronics/sanyo-eneloop.html. August 2008.
24. Reuters. January 23, 2007. Mitsubishi Heavy to make lithium ion car batteries. www.reuters.com/article/tnBasicIndustries-SP/iDUST272120070123
25. Boschert, S. 2007. *Plug-in hybrids: The cars that will recharge America.* Gabriola Island, BC: New Society Publishers.
26. Abuelsamid, S. December 6, 2006. Cobasys providing NiMH batteries for Saturn Aura hybrid. Autobloggreen.com/2006/12/05/cobasys-providing-nimh-batteries-for-Saturn-Aura-hybrid/
27. http://www.seqway.com/personal-transporter/lithium_ion.html
28. http://www.sciencemag.org/cqi/content/abstract/1122716
29. Buchmann, I. 2006. Charging lithium-ion batteries. Cadex Electronics Inc. BatteryUniversity.com
30. www.carb.ca.gov/

Index